GUIDE TO OUR GALAXY

For more information and articles visit our website.

www.planetseekers.com

Please review this book on Amazon reviews mean everything to me.

TABLE OF CONTENTS

Introduction 1

Chapter 1: The Pale Blue Dot 2

    The Anthropic Principle 2

    An Evolving Planet 3

    A Shooting Gallery Of Billiards 3

    Earth's Composition 4

    A Restless Planet 5

    Earth's Habitability 7

    Usable Real Estate 8

    Earth's Future 9

    The Perfect Earth 10

Chapter 2: Evolution, Civilization & The Quest To Explore 12

    The First Civilizations 14

    Sumer 14

    Ancient Egypt 15

    Norte Chico 15

    Olmec 16

    Indus Valley 16

    China 17

    Beyond The First Civilizations 18

    Rise Of Islam And A Great Change 19

    The Age Of Enlightenment 21

    The Great Technological Ladder 22

    Our Undetermined Future 25

The Cosmos As Our Savior 26

Chapter 3: Exploring The Cosmos 29

The Cities Of The Universe – Galaxies.. 30

Elliptical Galaxies – The Milky Way's Future 31

Irregular And S0 Galaxies – The Unwanted Offspring....................... 32

Stars – Life's Parents.. 33

The Birth Of A Planetary System 34

Star Classifications – Life's Sweet Spot... 34

O-, B-, And A-Class Stars – Hot And Heavy 35

F-Type Stars – On The Edge 36

G- And K-Type Stars – The Sweet Spot 36

M-Type Stars – An Enigma 37

The Stellar Graveyard 38

Habitable Zones.. 39

Venus And Mars – A Unique Family 40

A Galactic Habitable Zone 42

Alien Planetary Systems 43

The Golden Age Of Planet Hunting... 44

Detection Techniques 45

A New Generation Of Telescopes 47

What Will We Find?... 48

Chapter 4: The Boundaries Of Habitability............................ 49

Introducing Super-Earths... 50

Gravity Of Super-Earths... 50

Geology Of Super-Earths 51

Atmospheres Of Super-Earths 51

- *Super-Earths Are Everywhere!* 52
- *Pushing The Boundaries Of Earth-Like Planets* 53
- *An M-Dwarf's Younger Years* 53
- *Tidally Locked Worlds* 55
- *Habitability Of Tidally Locked Worlds* 56
- *Movie Example – White Dwarf* 57
- *Just A Little Instability* 57

Other, Even More Exotic Worlds 58
- *Habitable Moons* 58
- *Rogue Planets* 59
- *A Collective Evolution* 61

Chapter 5: Possibilities Of Alien Life 63
- *The Drake Equation* 63

Detecting Another Civilization 68
- *Sending A Signal* 68
- *Detecting A Signal* 69
- *Atmospheric Signatures* 70
- *The Fermi Paradox – Where Is Everybody?* 71
- *The Detection Conundrum* 73
- *The Existence Conundrum* 74
- *The Great Filter* 75
- *Highlighting Likely Filters* 76
- *Technological Self-Destruction* 76

Our Successor: Artificial Intelligence 77
- *Fatigue And Lack Of Motivation* 80
- *The Vastness Of Space And Time* 81

 Making It Through The Great Filter 82

 Contact Through Domination 83

 Contact Through Communication 83

 New Equations Are Needed 84

Chapter 7: One Among Many **89**

Further Reading **92**

INTRODUCTION

I have always held the belief that science is fueled by curiosity. My interest in the cosmos, and consecutively of our place in this tiny inconsequential planet called Earth, was borne from my personal quest to understand what really lay outside of our beautiful world and our true place in this pale blue dot. I felt there was an innate need to address some of the fundamental questions all of us have about the universe and do so in a manner that everyone can comprehend regardless of their proficiency in science.

If you've ever wondered about how we came to be or how our complex civilizations developed or yet if there are other planets that can harbor life, this book makes an honest attempt at answering all of those questions and more in a way that is fun and yet thought provoking.

From the dawn of time, humans have aspired to venture out to the vastness of space. People through the ages have looked up and wondered if we indeed are alone. Our science tells us that the basic elements of life are abundant in nature which means life can well and truly be developing in one or perhaps a number of other planetary systems simultaneously. However, our practical observation tells us that while the basics of life can be found easily, the perfect hospitable environment that is needed for life to foster and prosper is quite rare which makes the Earth so very special. But we must remember that there are billions of galaxies out there with multiple trillion star systems and who knows, maybe one of them would have conditions similar to earth!

Humanity must not lose its motivation and resolve to voyage into space. This book is a small effort in awakening the scientist in you, to energize you, excite you to be awed by our beautiful universe. Keep looking up!

CHAPTER 1: THE PALE BLUE DOT

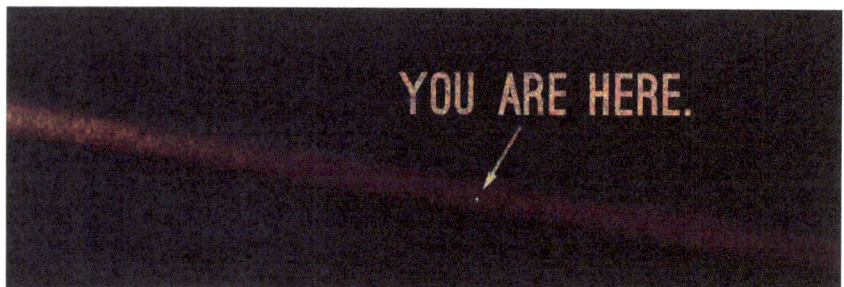

A sunny weekend has arrived, so you take your family on a road trip to some distant countryside cabin. The breeze whips through your hair, and a warmth saturates your skin from the Sun's rays that pierce the partly cloudy sky. The road follows a stream as your family glimpses a deer licking up water along the stream's edge. A fish jumps out of the water briefly. A side road through a dense forest takes you up a winding climb for another hour before you reach the cabin, nestled along a pristine glacial lake. As you get out of the car, you can hear birds singing in the trees and the croaking of frogs at the lake's edge. What makes this scene so beautiful? It appeals to us because the planet is not only suited well for life, but life is suited well for the planet. That is to say that over the billions of years of evolutionary history, both Earth and life have changed to become more supportive of and compatible with each other. A blue sky and crystal clear lake surrounded by forests is beautiful to us because we have evolved to view it that way. It is now instinctual for most people to see beauty in nature. To be in awe of nature leads humans to treat nature with respect, which helps with our survival.

The Anthropic Principle

The relationship of Earth and its life may apply to other planets. There may be life in every system we point our telescopes toward. The pessimistic view, on the other hand, is that Earth is the only planet able to support life in the entire Universe. While we have limited knowledge of what conditions life can adapt to, we do know that in our own solar system, Earth is exceptional. Our planet certainly appears fine-tuned for life when compared to every other planet, and scientists have debated how this came to be. One idea for why life exists at all is the anthropic principle, proposed by Brandon Carter in 1974. The principle states that because life requires a very specific set of laws and conditions for it to appear, the Universe to which life finds itself must by default be compatible with supporting life, otherwise life would not have arisen in the first place. A stronger view of the principle states that only life bearing universes would ever appear anyway, and there is no alternative for life to be selected for or against.

Earth may also apply to the anthropic principle in its natural fit for life. The conclusion then is that there should be other Earth-like worlds in the Universe, even though we have yet to discover, or at least confirm, any of them. Although the ever increasing count of planets in other star systems indicate that our being exceptional is highly unlikely, it is still possible that Earth is the only place in the Universe to support life. This possibility would suggest that the anthropic principle is incorrect, and Earth and its abundance of life is a fluke. That we have detected neither lower life forms nor more evolved life forms on other worlds can also be viewed as supporting this theory. What an epic and humbling lottery we have won indeed if our planet is the only place with life in the Universe!

An evolving planet

Earth was created in six days, and on the seventh day God became tired and needed to rest… wait, that's not the right story. Earth actually formed 4.54 billion years ago when the solar system coalesced from a collapsing cloud of gas and dust.

A shooting gallery of billiards

For the first few million years after its birth, our solar system was a chaotic place. Collisions of objects was a non-stop affair. Just as Earth's molten surface was solidifying, a Mars-sized object collided with our planet, and some of the debris coalesced to form the moon. The remaining debris eventually fell back to the planet.

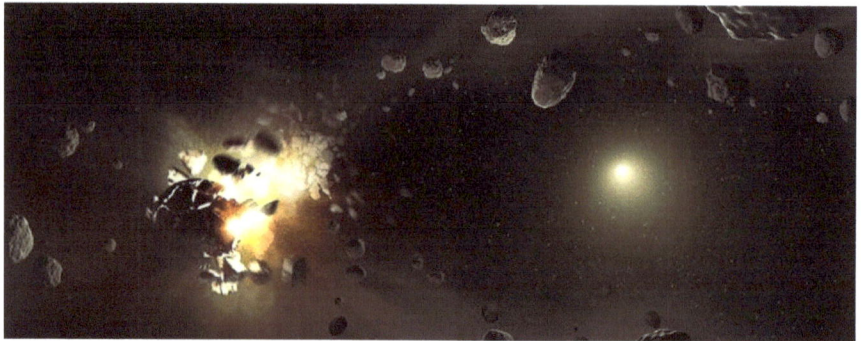

Millions of years later there was another series of impact events known as the Late Heavy Bombardment (LHB). This second wave is thought to have been caused by a shift in the orbits of the two gas giants, Jupiter and Saturn. As these planets migrated, at times toward the Sun and at times away from it, they knocked around the smaller inner planets so much that any primordial planets within the orbit of Mercury were sent careening into the Sun, or ejected out of the solar system entirely. If this had not happened, the solar system could very well have more than a dozen planets, including several larger versions of Earth-like worlds called super-Earths.

The chaotic time of the LHB helped Earth to develop its current life sustaining conditions. The rain of comets and meteors, rich with water, vaporized upon impact, depositing on our planet complex organic compounds. If the Late Heavy Bombardment had not occurred, or came much later, Earth might only have a fraction of the water it does now. One isolated ocean or an endless landscape of large lakes may have formed instead.

Earth's composition

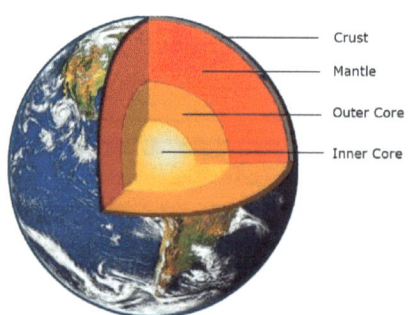

From the core of our planet to the atmosphere surrounding it, each layer plays a critical role in creating and maintaining conditions suitable for life. After the Mars-sized body hit Earth, the heaviest of materials were absorbed into the deep bowels of the planet, including precious metals like silver, gold and platinum. Nearly all of these elements sank right to Earth's core, thousands of kilometers below the surface (and completely out of reach of any possible mining operation).

The remaining elements formed the outer core, mantle and crust. If life was ever to emerge on Earth, these lighter elements would be required. Earth later obtained additional heavy elements from the bombardment of comets and asteroids. Earth's veneer of asteroid deposited heavy metals would later be crucial for powering human civilization. Most of the electronic devices we use today have small amounts of a group of these metals called rare earth elements (REEs).

There are more than a dozen rare earth elements mixed throughout the planet's crust, such as terbium, used in televisions, neodymium, used in hybrid automobiles, and thulium, used in high-efficiency lasers. Interestingly, many REEs are actually quite abundant. What gives them their name is the difficulty in separating the metals from other surrounding metals.

A restless planet

Because the temperature of the crust is so much cooler than at the core, the inner material convects. Convection is the process that heats your food evenly in the oven. The coils at the bottom heat the air, causing the air to rise and then cool at the top, where it falls down the sides and back to the coils. Convection in the Earth moves material from the crust down into the mantle, and eventually back up again via volcanoes. This movement also gives rise to the planetary magnetic field. Just like how we breathe in oxygen and exhale carbon dioxide, Earth also has a natural capacity to absorb and release gases. Carbon dioxide is a powerful greenhouse gas that would cause Earth to overheat if there was no way to regulate how much CO_2 is in the atmosphere. The planet absorbs carbon dioxide out of the atmosphere when rocks and plant growth settle on the ocean bottom, where the carbon dioxide becomes baked into the crust. The heavier oceanic basalt crust is eventually pulled under the continents and into the mantle. The material melts and safely releases the stored carbon dioxide under kilometers of molten rock.

The process of carbon capture takes a minimum of hundreds of years to occur, so the current carbon dioxide in the atmosphere will likely remain for the rest of our civilization's near-term existence, unless we can artificially pull that carbon dioxide back out of the atmosphere through more expedient (and safe) means. Earth is always changing. The planet's plates are constantly in a state of movement. While a few centimeters per year might not seem like a lot, over 100 million years this will shift a continent 3,000 kilometers! Pent up forces over decades will cause a sudden release of energy, creating earthquakes and the deformation of nearby land areas. As the continents push against each other, they will build up huge mountain ranges like the beautiful Rocky Mountains, islands like the Hawaiian Islands, and meandering rivers like the Nile River Delta and the great Amazon.

In addition to the continents, the oceans help the Earth to actively regulate itself. A significant amount of water is being warmed in some areas and cooled in others, constantly generating mighty storms. These storms begin as small, low-pressure systems in the warmer latitudes, quickly grow and migrate to cooler areas, where they eventually reach land and cause storm surges. Regular El Niño and La Niña events over the oceans affect entire hemispheres with altered wind and rainfall patterns, resulting in severe droughts like those in California, or mudslides in Leyte Island, Philippines, from too much rainfall in a given season.

Earth is also constantly moving through space, revolving around the Sun once per year and once around the galaxy every several million years. The planet also shifts on its axis slightly over thousands of years, which causes long-term seasonal changes that in turn affect the survival of life. The shifts in axis can also disrupt civilizations with drought, as was the case with the Mayans, or provide a burst of sustenance with decades of plentiful rainfall.

Shifts in axis can also cause or exacerbate ice ages. During the last ice age, sea levels dropped by dozens of meters as oceans became completely covered in surface ice. This resulted in a land bridge between Asia and America that our ancestors discovered just as they were entering Siberia. Bands of people slowly migrated from Asia to America over the land bridge for thousands of years before the bridge submerged again in newly rising seas.

Earth's habitability

Water is the first thing to consider when gauging a planet's habitability. On Earth, life revolves primarily around water. Expansive oceans help to regulate atmospheric temperatures and ensure a robust hydrologic cycle. There is not too much water though that would result in a water world (a world covered 100% by a single ocean). Water covers 71% of Earth's surface, leaving enough land area available for life to evolve into more complex forms, including a form that would one day end up building a technological civilization.

Earth's atmosphere also plays a major role in making the planet hospitable to life. The atmosphere is composed of 78% Nitrogen, 21% Oxygen, and other trace gases. Oxygen levels in the atmosphere have fluctuated in the past, from trace amounts at nearly 0% to as high as 35%. Oxygen also made – and maintains – the ozone layer, which is critical to shielding life from harmful UV radiation from the Sun. The rays that do get through can cause a severe sunburn, or worse, a facial wrinkle.

Gravity is another habitability requirement that keeps everything together. Gravity keeps our feet on the ground as the planet spins at 1,670 kilometers per hour. Earth would have to spin about ten times faster for objects to be ejected into orbit. Humans could technically survive without gravity, as we do on the international space station, but there would be serious side effects. The constant downward stress exerted on our bones keeps them from losing density. Intracranial pressure of the head is also kept in check, preventing long-term eye damage. Everything from eating, digestion, and excretion are assisted by gravity.

Our moon is one of the biggest that are observable in the solar system. Having a large moon helps our planet to maintain its tilt toward the Sun. Earth's tilt is currently at 23.4 degrees, resulting in relatively mild changes in the seasons. Without a large enough moon to stabilize the tilt, climate shifts would cause the most severe of ice ages to occur every few thousand years, instead of every few hundred thousand years. Each ice age would be more extreme as well. Warming trends would have a similarly exaggerated effect.

The rotational rate of the Earth is important in determining surface wind speeds, and to some extent assists the oceans in maintaining an overall mild variance in global temperature. Our planet today has a pleasant average wind speed of about 10 kilometers per hour. This is fast enough to scatter seeds and other particles necessary for reproduction, but it is not so fast that trees are uprooted. When Earth was less than a billion years old and a day was just 12 hours long, winds were so extreme that standing upright would have been difficult, if not impossible.

The properties of Earth's inner material may be the most important part of the planet. Earth's active interior creates a fluidic dynamo – the physical motion of material – in the planet's outer core, which generates a globally encompassing magnetic field. The field not only protects life like the atmospheric ozone layer does, but it also keeps the solar wind at bay, thus preventing the atmosphere from being eroded. Without an atmosphere, the oceans would quickly sublimate into space.

Usable real estate

Earth may be a life-rich planet, but it is still carefully balanced on a knife's edge in terms of its ability to support life. In fact, the majority of its surface is uninhabitable. Getting a tan on the beaches of Bora Bora or El Nido may make us feel like we live in a global paradise, but a quick trip to chilly Antarctica, or the driest non-polar place on the planet, the Atacama Desert, we would quickly realize that not all of this world is suited to humans. As the saying goes in the real estate business, everything is location, location, location. Just 15% of Earth's land is worth occupying, whether to extract resources or to settle. Very few would build a cabin in the middle of the Atacama Desert, or on the ice shelves of Antarctica.

While most of the Earth is not fit for human settlement, the remaining total area that is fit for our use is still generous. Even though human population is now more than 7 billion persons, there is enough area to build (though not necessarily sustain) an advanced technological civilization for perhaps as many as 15-20 billion. Still, we are ruining much of the available land, and may soon reach the point where it is insufficient not because of overcrowding, but because of mismanagement. Regardless of a planet's size, overcrowding and resource management will likely be a consideration for any advanced civilization. Even the largest of super-Earth planets, with two times or more surface area than Earth has, could fill up with walking, talking beings at some point. By comparison, if your current 500-square meter home is full of stuff and you move into a new

1,000-square meter home, you will probably want to buy that bigger bed, larger dining table, have another child, and so on.

With so little of Earth's land being useful to humans, it doesn't help that the majority of the surface of our planet is covered by water we cannot drink. The salinity of seawater makes it unusable for crop irrigation and human consumption; in fact, less than 1% of all water on the planet can be used for these purposes. If you could gather the entire quantity of freshwater both on Earth's surface and any underground, the volume would amount to a sphere with a diameter of 273 kilometers, or the distance you would travel driving at 92 kilometers per hour for three hours.

Earth's future

Earth has entered the 'Anthropocene', a geologically distinct epoch where human activity has forever made a mark in the geological record. Future alien archaeologists will be able to find this mark, should they ever visit and we are long gone. Although our planet is always changing and entering new epochs (usually over millions of years for each epoch), and will continue to do so regardless of human activities, eventually it will change permanently for the worse. We can observe other planetary systems to get a peek at what is in store.

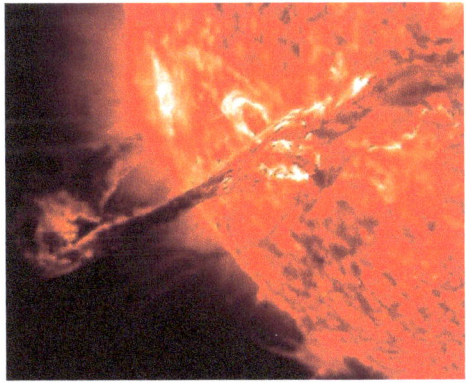

Here is how it is going to play out for Earth in the distant future with an increasingly elderly Sun baking its surface: during the next four billion years the Sun's temperature is going to continue to slowly increase, rising by another 10% or so. That doesn't sound so bad, until we realize that Earth is already on the inner edge of the solar system's habitable zone. As the Sun's temperature increases, this zone will move further out. While Earth will not become the next Venus, with surface temperatures that can melt lead, the oceans will still boil off.

Earth's surface will be baked sterile. The upper layers of the atmosphere will be carried away by the solar wind, and any remaining water vapor will seep into space. Eventually, our planet will lose its entire atmosphere. One can take heart, though, in knowing that life has at least a few hundred million years before the Sun will start to roast us.

The perfect earth

Earth is overall a great place for life. At first glance it may seem like all of the knobs are set perfectly to meet our needs -- water to quench our thirst, food to fuel our bodies, and oxygen to breathe are all seemingly available in abundance. Humans have settled everywhere there is land and these three needs are met, and flourished, so what's there not to like?

The knobs of perfection are in fact quite a bit off from what planetaryscientists can envision, as well as by anyone who takes a moment to think about places on Earth that are not habitable. What we know about the limits of chemistry, geology, and other sciences would suggest that there ought to be planets out there even more suited to life than Earth.

If you had control of the planetary construction dials, what sort of improvements would you make to our planet in order to create a more habitable place for humans?

With more than a thousand confirmed planets in the galaxy so far, scientists are beginning to doubt that Earth is the best real estate in town. Our planet is certainly good, great, even fantastic for life, but not perfect. The Earth Similarity Index is a recently established measure of habitability. Hundreds of factors are taken into consideration, but most important are a planet's size, interior composition, atmospheric composition, presence of a magnetic field, and presence of liquid water. Earth scores a 1 on a scale of 0 to 1. Mercury is 0.86, and the moon is 0.56. While we haven't yet discovered any planets higher than 1, the scale does allow for this possibility with some of the suggested changes below.

Clean water and safe food are the first two things that come to my mind when thinking about what a perfect Earth would need in greater abundance. It is ironic that our planet has more water than land surface, and yet thousands of children die every single day from a lack of safe drinking water. Even when the water is not contaminated, if it is salty like that of Earth's oceans, drinking it will only cause one to dehydrate faster.

Hundreds more children are poisoned from wild or improperly prepared foods.

A higher level of oxygen in the atmosphere would help in fueling our bodies for very energy-intensive tasks. With more oxygen in the air, lung capacity would increase, requiring a smaller set of lungs. The reduction in lung size would free up the body's resources to focus on other operations, perhaps increasing the brain's efficiency, enhancing intelligence. It took Earth a billion years to build up enough oxygen to fuel multicellular life, so it's clearly a key component in making complex life happen.

More oxygen would come at a cost, however. There would be an increased chance of wildfires. The doubling of oxygen to about 40% could cause wildfires to burn out of control even during a heavy rainstorm. This amount of oxygen in the atmosphere was last seen about 300 million years ago. Since then, levels tapered off to the current 21%. Any lower than 15% and extinction events could occur, alongside an onset of sudden evolutionary changes in species in a struggle to adapt. We might also think that a perfect Earth would have no natural disasters, but as described above with atmosphere regulating plate tectonics, a planet cannot be unmoving and expect to support something as dynamic as life itself.

Understanding what life is, what makes our planet capable of supporting it, as well as life's limitations, will help us in the distant future when humanity looks toward space for another place to call home. Until we find such a world that is worthy of being called Earth 2, the worst thing we can do is shrug off the responsibility of taking care of our one and only home. That process of understanding begins with the origin of life itself.

CHAPTER 2: EVOLUTION, CIVILIZATION & THE QUEST TO EXPLORE

The Paleolithic era, or Early Stone Age, is the earliest period in which we can identify ancient humans as a distinct species from previous ancestors. During the Paleolithic era many Homo sapiens still lived alongside the Neanderthals and Denisovans, both close relatives, genetically speaking. There is even evidence of the three interbreeding. Some scientists believe that one reason the Neanderthals eventually disappeared is that the Homo sapiens population of the time was comparatively much larger, and the Neanderthals were simply absorbed into their genetic pool. Because of this cross-breeding, many of us have a little bit of Neanderthal in our DNA.

After a long and barbaric Stone Age that lasted millions of years, other ages that were relatively more civilized quickly followed. The Neolithic period came at the end of the Stone Age and it marked the beginning of humans settling and cultures forming. The Neolithic Revolution was also the transition from hunting and gathering to the use of agriculture. A side effect of humans roaming less and settling more was the encouragement of cooperation and exchange of ideas.

Language flourished, as well as the need for social stability and security. As society became more complex, the demands upon our brains increased. Agriculture helped to feed an energy-hungry brain that from generation to generation evolved in complexity. The very concept of progress and advancement was also born around this time. Once the secrets to creating a stable and productive society were discovered,vast amounts of time were freed up. New leisure activities like art,music, and science developed. For the first time in all of history, a creature on the planet had the capabilities to craft its own future, and even have fun in doing so.

The Neolithic Revolution was a time when each new discovery was changing the world. The concept of schools had yet to come about, but many cultures still shared resources and processing techniques. Regional knowledge expanded and, over time, much knowledge became global. The wheel is an excellent example of an early tool that improved the efficiency of countless human pursuits. When discoveries and inventions transformed the civilizations, a new technological age was ushered in. The next to follow were the Bronze and Iron Ages.

The Bronze Age is so named because it was the era when we discovered how to make bronze alloy. Bronze is a mix of copper, tin and a variety of other metals, such as zinc and nickel. When heated and melded properly, extremely durable tools can be made. The alloy marked the first time we could build things with materials beyond the raw resources that were cut down, dug up, sculpted, etc. Bronze was also an essential component in some structures, including the first plumbing system in the Indus Valley Civilization around 2700 B.C.

After the Bronze Age, we learned how to make even stronger metals. Iron (and later its steel alloy) was first forged at the start of the Iron Age. Iron was not discovered previously because it was so difficult to mine – an even more arduous process than figuring out the right metal mix to make bronze. The process of mining was also a new concept. As mining spread from village to village, blacksmiths quickly popped up. Iron was used to create a wide variety of items, especially deadly weapons. Bronze swords, in contrast, just didn't cut it on the battlefield.

The Iron Age also marked the appearance of the first true alphabet, which came from the Greeks. Unlike Greek's precursor, the Phoenicians, letters represented both consonants and vowels, as opposed to only consonants. The Greeks were mighty traders of the Mediterranean Sea during the Iron Age. Travel and trade was a way of life. In order to keep track of everything they traded within such a diverse set of lands and associated languages, they invented an alphabetical system of writing in order to make record of transactions. Alphabets became a standard of nearly every civilization thereafter.

The first civilizations

Many civilizations changed the course of history for humanity. There are six great civilizations recognized as being among the first: Sumer (Mesopotamia), Ancient Egypt, Norte Chico (Mexico), Olmec (Mexico), Indus Valley (Pakistan), and China. While only a couple have survived the test of time (Egypt and China), each of them contributed important elements to the development of today's modern civilizations that are now capable of reaching for the stars.

Sumer

The first civilization to appear in southern Mesopotamia was in Sumer, which is known as the "cradle of civilization." The Sumercivilization began around 4000 B.C. and lay between the Tigris andEuphrates rivers, in today's Iraq and Kuwait. The Sumer built a wide array of cities, which were the first examples of urban life. Walls were built to keep out invaders, which worked for at least a few hundred years. Unfortunately, the main wall (about 250 kilometers in length) was not well defended, and there was nothing to stop the hordes from simply walking around the sides to pillage the towns and cities.

Sumerians invented many things, including the wheel, the plow, and glass. The first inklings of organized religion started in Sumer. One reason that the Bible can be interpreted to suggest Earth's age at a scant six thousand years is that the first drawings and writings we know of were created six thousand years before Christ, in Mesopotamia.

Ancient Egypt

Around the time that Sumer sprouted up, more than a thousand kilometers to the west, Ancient Egypt flourished along the Nile River Delta. While there were smaller groups in the area before Egypt was founded, it took the first great pharaoh to unite them all into one civilization. Advancements in agriculture played a big part in the lasting power of Egypt. The civilization did extremely well in cultivating arid land, as well as trading with neighboring city-states to build a healthy economy and prosperity for its people. They were obviously very talented at construction. Egyptians not only built the Great Pyramids, but also many temples and obelisks that marked their territory for hundreds of kilometers.

Thousands of years would pass from one slow stage of technological development to the next, starting in the Bronze Age and culminating into a great civilization, declining through war and other problems, eventually changing into what we see as modern Egypt today.

Norte Chico

25,000 years ago, there was increasing glaciation that yielded an ice age which lowered sea levels. A new land bridge, the Bering Land Bridge, appeared between what we now call Eastern Siberia and North America. The Eastern Siberians achieved the incredible feat of trekking for thousands of kilometers over rugged mountains, in freezing temperatures, facing countless dangers. This migration occurred over the course of thousands of years. These ancestors from Eastern

Siberia colonized North America and eventually became known as the North American Indians.

Some tribes continued traveling south into the South American continent. Around 4000 B.C., many settled in the area along the coast of what is now north-central Peru – this became the Norte Chico civilization, which emerged about a thousand years after the Sumer civilization. Initially, the Norte Chico people had virtually no art or use of symbolism. They were a very peaceful people, as no evidence of the use of weaponry has ever been found. The grandest of the Norte Chico's achievements were their many monuments, such as their platform mounds and step pyramids, made possible by advanced ceramic techniques. There is also evidence that a rather complex government was in place for many hundreds of years to manage the diverse population.

The demise of the Norte Chico came about slowly over hundreds of years through migration of people from outside the civilization. The knowledge and ceramic techniques the Norte Chico developed were taken to other lands by groups that quickly grew into their own nearby settlements. A thousand years would pass before another great civilization was built in the area, the Chavin, and eventually the Olmec, eclipsing the original Norte Chico people.

Olmec

The Olmec came to the area around 1200 B.C., thousands of years later than the Norte Chico, though they were the first to settle in the northern Mexico area. They reigned in the Mesoamerica region for a couple of thousand years before mysteriously dying out. Historians are not certain what caused the decline, but it is highly likely that it was caused by drought, or perhaps an internal unrest of some kind.

Many of the Olmec's great works and achievements would end up being used by hundreds of civilizations that followed, including many techniques for crafting beautiful sculptures. Their culture included complex religious and creative institutions which flourished over a wide area. The advent of the Mesoamerican Long Count calendar, writing symbols, and a variety of sports and games (including a primitive form of football or "soccer") was thanks to the Olmec. Sadly, many of the Olmec's works and cultural artifacts were pilfered or destroyed during the Spanish Conquest in the sixteenth century.

Indus Valley

The Indus Valley civilization emerged at the start of the Bronze Age in the area that is now Pakistan and northwest India. The Indus Valley civilization was thriving at the same time as the Sumer and Ancient Egypt, but it was the largest of the three. Metallurgy was their specialty, including the ability to construct multi-story buildings, plumbing systems, and extensive irrigation works. Many of these techniques would later be used by the Roman Empire, and then they were resurrected by modern civilizations.

A likely cause of so many advances in technology in the Indus Valley was the density of people in the cities. As more people were forced together, society had to come up with new ways to handle the problems that came with a crowded living space. Ways to feed the people required innovative uses of wood and metals to irrigate and farm the land. Nearby city-states would often attack to gain control of these consolidated resources, so advanced weapons and defensive structures had to be developed, as well as laws for internal peace and order. Numerous religions formed that consoled a population that was constantly in a state of uncertain change.

China

China could be said to be the greatest civilization that has ever existed for multiple reasons, including its sheer longevity and also its technological prowess. Ancient China, like other civilizations, had to defend against invasion. To prevent roaming Mongol hordes from taking over their lands, the Chinese built the Great Wall of China over the course of many centuries. With such a large empire, it is a testament to the strength of the Chinese that they managed to keep their civilization intact to this day. Only the Mongolian Empire was able to secure a large piece of China's northern territory. Chinese civilization today stretches from the Tibetan plateau to the northeastern Mandarin region, and far southeast to the

Cantonese villages.

The Ancient Chinese were the first to invent paper, the compass,printing, and gunpowder. Interestingly, the Chinese originally used gunpowder to create fireworks displays for children. It was not until an accident at a storage facility that gunpowder revealed its true potential, both in its destructive capacity, as well as its excavation potential.

The Ancient Chinese also developed music theory with a unique five-tone scale known as the pentatonic scale, in contrast to today's western diatonic/heptatonic 7-note scale. Much of the music theory of Ancient China is otherwise quite similar to the rest of music theory taught in schools around the world today. Their most unique instruments of the time period were made of bamboo, an excellent resonating material. Flutes and reeds were highly popular instruments. Their two distinct languages, Mandarin and Cantonese, also have a four-tone and five-tone differences, respectively.

Beyond the first civilizations

Many other civilizations would sprout up after these first ones appeared. Often, new civilizations would appear after a weak one would split from war and attrition, disease and famine, or changes in climate forced some of its inhabitants to migrate to other areas. Therise and fall of civilizations occurred particularly in Europe, until the Roman Empire around 27 B.C. began to unite all of the surrounding warring tribes and independent villages.

From the Atlantic Ocean to the Caspian Sea, the Roman Empire grew to be the greatest civilization the world would see for hundreds of years after its existence. While all civilizations that came before achieved great things, the Roman Empire reached new heights in construction techniques with expansive cities and road networks, the most complex language to have ever existed, Latin, an alphabet still used today, and a great many systems of law. If it were not for barbarian hordes attacking, and internal unrest caused by the empire attempting to expand beyond its capabilities, it'squite possible the Roman Empire would have been the first civilization to reach what we would define as a modern society. As with all of the civilizations that came before, though, the Roman Empire eventually began to fracture around 117 A.D. with the death of the emperor Trajan.

The Dark Ages began shortly after the Roman Empire collapsed. Progress in resource discovery and economic growth came to a halt in Europe. Centuries passed with human society using nearly the same materials and resources they had used during the previous ages. In fact, nearly every natural metal that exists today, not counting alloys, was discovered before the Middle Ages even began. The refining process and discovery of new alloys was only picked up again around the start of the Enlightenment period in the 17th century.

Rise of Islam and a great change

A wide array of differing beliefs, traditions, and rituals arose within these civilizations, including the three most widespread monotheistic religions: Judaism, Christianity, and Islam. Each has transformed the world in different ways, yet the Golden Age of Islam was unique in that it once provided a foundation for scientific study.

Islam originated back in the early seventh century in the Arabian city of Mecca when the prophet Muhammad claimed to have received a revelation from God (Allah). Muhammad was alone in a cave at the time he had the revelation. He shared the story of what he had seen with others, and the religion of Islam began to take root. The story of someone alone in a cave receiving a vision has been told in various ways in many other religions. Stories are powerful tools for directing a society toward a set of goals, regardless of whether the underlying assertions are true or not. Muhammad single-handedly managed to sow the seeds of a worldwide religion with his tales and conquests.

The following centuries were a time of great progress for the Islamic civilization. One extraordinary achievement was a vast mercantile trade network that stretched all the way to China with their sophisticated three-masted caravel ships. Roads were constructed, along which many villages were built, and different kinds of food were grown, creating a variety of goods and wares for sale. Astronomy and mathematics became a significant public scientific endeavor. Many of the stars in the sky and the constellations have names that were given to them at this time. Muhammad Algoritmi, whose Latin name means algorithm, was a Persian mathematician who helped spread advanced mathematics, including algebra, to the West.

These innovations contributed to the Golden Age of Islam which lasted from around the eighth to the thirteenth centuries A.D. One of the greatest ideas in the religion was the ijtihad, which means independent reasoning. Individuals were taught to think for themselves  and to question the beliefs and ideas of others. Few things were out of the bounds of discussion, with the exception of religious texts and laws. Ijtihad focused on the betterment of the whole civilization through the work of individuals. The basis for today's scientific principles was founded on some of the early versions of ijtihad. Taqlid, on the other hand, is an idea in Islam that focused on the individual's obligations to the religion – and it demands unquestioning acceptance. Taqlid is about conditioning oneself to Allah, and not asking why. Elements of Taqlid would eventually find their way into the Koran, and they are still practiced by Muslims today. Questioning the ideology is blasphemous and punishable through a variety of laws.

Even verifying facts for scientific purposes is often not allowed. The anti-rationalist school called Ash'ari arose partly because some people were so opposed to scientific reasoning. This anti-science movement grew in popularity and even resulted in the banning of teaching of manysubjects in schools. As Taqlid continued to outlaw science and criticalthinking, progress gave way to decline. War, pillaging, suffering, and disease contributed to a steady unraveling of the Golden Age of Islam. Greed within the Islamic empire slowly took its toll; great libraries and institutions lost support or were completely destroyed, and the overall quality of life and opportunities of the people waned. A vicious cycle of corruption quickly set in which exists to this day. Unfortunately, as time went on, taqlid triumphed over ijtihad. Taqlid appealed to those who had true power, as well as to those who wished to maintain the illusion of power. Ideas can be powerful motivators for good, but they can also be motivators for poor decision making, especially when they are not based on facts and reason. Once the process of forbidding critical thought begins, history shows that it can often take a long time to recover from the strife and retrogression that causes a spiral into an age of darkness.

From the start of the Bronze Age to the end of the Middle Ages there is a span of at least 4,000 years. Extend that back by another 5,000 years to include the Neolithic period and its first civilizations, and roughly 10,000 years of civilization is on record. Before civilizations even appeared, Homo sapiens existed for more than a hundred thousand years.

The age of enlightenment

The Age of Enlightenment was an era of intellectual advancement that flourished from the 1680s through the 1790s. There was a collective goal of exploration and progress in many disciplines, including philosophy, political thought, social theory, science, art, economics and law. Numerous closely knit nations cooperated freely in the exchange of ideas. Innovative thinkers of the period came from many countries, but the majority hailed from England, Scotland, France, Germany, and America. Traveling from one region to another had become safer and easier, which aided the spread of ideas.

During the Age of Enlightenment many great intellectuals rose to international prominence. Some of the most well-known are John Locke, Sir Isaac Newton, Thomas Paine, Voltaire, Thomas Jefferson, Immanuel Kant, Jean-Jacques Rousseau, Montesquieu, Benjamin Franklin, Adam Smith, and Mozart, among many others. What they had in common was a resolve to improve the human condition by understanding the world from evidence-based viewpoints that relied on logic and reason for discovery. Debates that occurred were very different from those of previous centuries that focused heavily on prejudice and superstition. The Enlightenment marked a turning point in history when humans committed to uncovering, proving and valuing truths, using such means as empirical examination, fervent deliberation, inner reflection, logical argumentation, and the scientific method.

One fundamental tenet of the Enlightenment was that individuals have inalienable rights that they are born with. As outlined by the Founding Fathers of the United States of America, the government's responsibility is to protect those rights; rights could not be considered as granted to individuals by governments, because then governments would have the power to take those rights away. Individuals were encouraged to know freedom, prosperity, comfort and happiness.

Individuals had freedom to practice their various religions, without interference from the state. Ending religious fanaticism that led to persecution and torture was an imperative to Locke, Voltaire, Jefferson, and nearly all others. Individuals were similarly allowed to compete in a free, laissez-faire market economy that would no longer favor the privileged as a matter of course – on the contrary, equal opportunity and social mobility began to permeate all elements of society. The common man was thus liberated, and the predominance of despotism began to peter out.

Just as important as having great thinkers to come up with ideas was a way to retain that knowledge and encourage its use in the future. Books were published in greater number than ever before, encyclopediae and dictionaries were compiled, and great libraries and universities were built to pass the knowledge and wisdom on to future scholars. Scientists understood how critical it was to remove biases that would otherwise distort the facts. Superstition was thrown aside, and by replacing it with truth, countless discoveries and inventions followed that brought the whole of society the opportunity to learn and discover. Schools became accessible to more than just the aristocracy. Healthful innovations like new medicines, improved hygiene, and safer food storage further improved the quality of life and increased the average lifespan.The Enlightenment was the greatest catalyst for progress humanityhad ever known, and its effects are still felt today.

The great technological ladder

Until the invention of mechanization, the typical daily life of a family consisted of someone managing the home, while others would work on the farm or go out hunting to collect food. Each generation would repeat necessary tasks day in and day out with little change. Today, we take for granted the technology that makes our daily lives easier, and perhaps arguably, more enjoyable. We lose sight of the many steps it took to get civilization this far. For instance, you may have gone to the kitchen to get something out of the fridge this morning. For that fridge to exist at all, a series of increasingly complex technologies had to come about first. They start with the metal housing, electrical conductivity to power its functions, and rare gases to drive the coolant. There are also the lights, rubber seals, and a nice colored finish.

You probably switched on a light within the past few hours. A set of technologies had to already be in place in order to get electricity to the bulb. There is the wiring within the building's walls, as well as transmission lines crisscrossing throughout the area, travelling perhaps hundreds of kilometers from a distant power station. That station is processing fuel that was sourced from afar, possibly on another continent. Even the power plant itself is a bundle of technologies with millions of components, requiring years of operator expertise.

Those not versed in construction techniques could still outline the resources and tools it would take to build a treehouse. Even though the details of how to make a comfortable, sturdy one might be unknown, people would have a basic idea of what was required. The details we would not necessarily know could be called "Known Unknowns." We know something is missing, and have a guess as to what it might be. Then there are "Unknown Unknowns," which we know nothing about.

This process of discovery and invention over many centuries in a stable environment will apply to every alien civilization out there as well, including the alien rescuers that might help us out come doomsday. A technological civilization like ours depends upon many materials and forms of energy to sustain itself and further develop. It is imperative that we keep innovating if we want to have a shot at surviving as a species in the long run.

Humanity has traveled a complex and unpredictable road to get toits current technological era. The price that we have paid is our presentreliance on technology in an ever more global fashion. The great kingsof long gone civilizations have been replaced by multinationalcorporations that are arguably more powerful than national leaders.Purchasing food from the neighboring farms has largely been replacedby trips to the supermarket to buy food that has been shipped acrossthe country, or even from around the world.Humanity's achievements have resulted in great feats of engineering like traveling to the moon. Other achievements have also resulted in great destruction, such as the nuclear reactor accident at Chernobyl, which caused tens of thousands of people to die of exposure to radiation. Countless times throughout history, civilizations have faced destruction, but in the last hundred years the pursuit of knowledge and technology has led us to the point that the destruction has the potential to be globally catastrophic. Technology is a double edged sword for humanity.

In addition to its complexity, today's civilization differs from those in the past in that its energy resources are dwindling. All of the easily accessible resources that a technological society needs are quickly being depleted. Even with technology that helps ease the transition to utilizing diverse, newer sources of energy, many resources will one day be permanently depleted (at least those on Earth). This is a problem that no pre-technological civilization had to the degree we see today.

As early as just a few centuries ago, if nearby drinking waters dried up, the population simply moved – a solution not suitable for millions of people today.

Over the course of just a few centuries, humanity managed to invent the technologies that allow us to understand what causes earthquakes, lightning, disease… we have uncovered the nature of the microscopic realm, and even have found ways to explore outer space. As grand as it is to behold, our civilization could be compared to a house of cards that may collapse as a result of any number of significant global disruptions. Think of the top levels of cards as the new upcoming technologies. The bottom row of cards are the raw energy resources we use to power the newer technologies. If the bottom row were to lose enough cards all at once, the entire house would collapse.

Civilization could be destroyed through numerous scenarios, including the mismanagement of natural resources, running out of nonrenewable energy sources like coal and oil, natural disasters, and even man-made pollution. In the event of a collapse, every technology that preceded the discovery and development of a future technology would need to be reacquired. If any of those resources are permanently depleted, including manpower, the house of cards we once had may be impossible to rebuild, at least of any degree that it currently exhibits.

Civilization urgently needs to find a way to live on other planets orbury its efforts and remain confined to Earth forever. The urgencyarises from two currently developing problems: overpopulation andresource depletion. Overpopulation means that too many persons areliving in an area which cannot support them. A trillion people could liveon Earth if the planet had enough resources to sustain them. As moreindividuals vie for increasingly scarce resources, conflict and controlover those resources is bound to occur.Everything from fossil fuels to rare metals are being depleted atfantastic rates. Many of them will run out within a generation or twobecause growing populations are using more of the resources. Thechallenge is to extract sufficient quantities of resources in space, and intandem make use of more abundant substitutes here on Earth in themeantime. If the technology can be developed fast enough, we have achance to not only relieve the problem of overpopulation, but alsoresource depletion.

The current population growth rates thankfully have a silver lining.Long-term population projections suggest that a leveling off of growthwill occur in the late 21st century. It is projected that more women willenter the workforce and will consequently have fewer offspring. Birthcontrol will also become more available. Also, as economic status andgeneral health increases for the population, aging parents will not needas many children to take care of them later in life.

Our undetermined future

If we truly wish to avoid going back to the stone age, we must be proactive in making sure of our safety as a civilization and the survival of our species. On the world scale, this will mean taking necessary precautions to protect our aging infrastructure like power grids from natural disasters or attacks. We must also do a better job at making sure food, water, medicine and the immediate necessities are at all times well stocked at the protected shelters in all major cities of the world. We must also work to improve and harden our military and civilian rescue equipment like radios, automobiles and electric generators. Countries also need to work together to prevent a global catastrophe.

If we fail now to reach outer space and tap the plentiful resourcesthere, humanity may be doomed to a limited existence on Earth,reduced to a patchwork of wandering groups in a hostile and energystarvedenvironment. Subsequently, we may plateau at a stage ofdevelopment much more primitive than our current one. Civilizationcould be set so far back that we would never be able to travel into outerspace again. Children in the new world would read about the old idealsthrough tattered history books, flipping past images of great cities androcket ships from what was once a grand civilization largely devoid ofthe problems they now faced.

Idiocracy (2006) is a satirical science fiction comedy movie set in adystopian future 500 years from present day. All of the modernconveniences are exaggerated, such as microwave meals, easyaccess to goods, and a sense of individual freedom and superiority.Society is seen to have long ago reached its peak of intellectualgreatness, now instead being pictured as a dumb society reliant on thesystem for literally every facet of life. Intellectualism is viewed as athreat, and technological progress has screeched to a halt. While lifecontinues on, it is evident that eventually the species itself is destinedto having a reduced intelligence, progressing within a handful ofgenerations to a stage not much smarter than the great apesthemselves.

Whether it's an immediate disaster, or the slow decline of ourintellectual capacity ('intellectual atrophy' as I like to call it), I personallydo not want to ever find myself in that kind of crumbling world. What atragedy it would be if civilization never recovered from a naturaldisaster or, worse, a disaster of our own making.One thing is certain if we ever had to rebuild-- for better or worse,the way of life for future generations would never be as it was for thosewho lost so much the first time around.

The Cosmos as Our Savior

Humanity is currently fighting to stay afloat on the great cosmicbarge that is the Earth. Achieving a permanent space presence wouldpresent a real jackpot: the resources we can harness from Near EarthObjects (NEOs) like asteroids and the moon. The process of colonizingspace can start as a business venture of mining asteroids. Even amedium-sized asteroid contains plenty of water that can be brokendown into hydrogen for fuel and oxygen for air, not to mention thetrillions of dollars' worth of rare metals. Eventually colonies could beestablished on Earth's moon, and thereafter on the moons of the outergas giants. All is conceivable with today's great thinkers working ongetting us to this point. Expanding our civilization's presence in the solar system – or maybe even throughout the galaxy – will not be easy or quick, just as it wasn't for the first explorers who crossed the Atlantic Ocean. The exciting part is that the possibilities are endless, just extremely challenging to get started.

While sentient creatures colonizing space may be a temporary blip in the evolution of the Universe, humanity has proven that it can be accomplished at least once. We should refuse to stall our trek to the stars when we haven't even left the proverbial driveway. Exploring the rest of our cosmic suburbia may prove to reveal only empty houses, but it's worth every effort because knowledge of the cosmos is empowering to humanity. As far as can be ascertained, we embody the Universe's best and perhaps only opportunity of leaving a legacy worthy of its existence.

We might one day learn how to prevent natural disasters from occurring. We might also one day learn how to push our planet into an orbit further from the Sun in order to escape the Sun's increasing heat. Moving the orbit of the planet may sound extreme, but it should be viewed as a very large engineering challenge to meet and not as an impossibility. The International Space Station (ISS) is one of many important stepping stones to reaching other locations in the solar system and beyond. All of the other planets and their orbiting moons will provide valuable construction real estate, as well as raw resources, to help encourage humanity to reach even further. These worlds will allow us to run experiments which would be impossible on Earth. Many of those experiments will be critical to ensuring the survival of future generations of space explorers that will not have the luxury of returning to Earth. For instance, it is extremely costly to build a telescope and launch it into space. Billions of dollars and years of preparation at a minimum is typically necessary, and there is still great risk of it blowing to smithereens upon launch. If permanent colonies were established in space, telescopes could be built directly there, and be far larger and more effective than anything we could ever launch from Earth.

Telescopes could be so massive, in fact that we could easily peer atthe atmospheres of millions of other worlds to see if they have life uponthem.At the very least, space will be impossible to overcrowd. There areenough rocky surface areas in the inner solar system alone totheoretically support trillions of humans and animals, not to mentioncountless locations for space stations.

If the public views space travel as not worth the risks, then we needto do better to educate the public. Human civilization's time to colonizespace is now, for we may not get a second chance. As we will explorein following chapters, this might be equally true for every buddingcivilization in the Universe. One shot to colonize space is perhaps allthat anyone ever gets.

CHAPTER 3: EXPLORING THE COSMOS

Our latest theories about the origin of the Universe suggest that, in the beginning, all of its energy was crammed into what is known as a singularity. Then suddenly everything, including time, sprang into existence in a "Big Bang." The Universe burst from the womb of nothingness for all of its future astronomers to detect. This initial birth period of the Universe is known as the Planck Epoch, and lasted about 10-43 seconds, or a tiny, tiny fraction of a second. Just as the Universe came into existence, the Planck era ended and another phase of expansion called "cosmic inflation" ballooned the Universe into a much larger size. At about 10-36 seconds, the Universe slowed its expansion as it began to cool. While cool by comparison to Planck era temperatures, the Universe was still at millions of degrees Celsius, and an even hotter place than the core of our Sun. At this time the four fundamental forces of nature – the strong and weak nuclear forces, electromagnetism, and gravity – separated out and began to act as they do today.

Microseconds later, as the Universe continued its relentless expansion and cooling phase, protons, electrons, and neutrons formed. The most fundamental building blocks of all matter now existed. After these first few chaotic microseconds of expansion, the Universe remained in a hot and dense state for the next 300,000 years. Unfortunately, there is no way to know what the Universe would have looked like to the human eye before this time. Everything was still so hot and dense, light itself as we can see it today had not yet come into existence. The Universe would have looked as opaque as a black hole to an outside observer. All of existence previous to about the 300,000- year mark is hidden from our telescopes.

At this young age of 300,000 years, elements heavier than helium could not be created. None of the elements that make up our bodiesexisted yet; in order for these elements to come into being, the hot andenergetic cores of stars would be needed to initiate a fusion reaction that could fuse together heavier elements. The first generation of stars formed a few million years after the birth of the Universe. They lived short lives of only a few million years, at most. When they died, they exploded, seeding nearby space with heavier elements, and providing the ingredients for the very first planets to form.

The cities of the universe – galaxies

A couple of hundred million years after the first stars formed, an intense bout of new star formation in clusters began to occur, forming the first galaxies. Galaxies can be thought of as the Universe's cities that drive the potential for life and civilization to blossom. Each galaxy contains anywhere from billions to trillions of stars, as well as countless gas and dust clouds. Many of these clouds are still in the process of forming new planetary systems.

While no two galaxies are identical, there are few enough differences to classify them into a handy set of categories.

Spiral galaxies are host to a large amount of stars, numbering into the hundreds of billions for the largest. They are shaped by gravity into a flat disc of stars circling around a core. This galactic core contains adense region of stars with the very center harboring a supermassive black hole. Extending out from the center are the arms of the galaxy, giving the galaxy its spiral shape. Currently, 20% of galaxies are spirals. Yet because they are so bright, 70% of the galaxies visible from Earth are spirals. The spiral shape of these galaxies is a bit deceiving. One might think that it is caused by the rotation of the galaxy itself, but that is not the case. All of the stars in a spiral galaxy revolve around the center in a slightly elliptical pattern. This pattern causes gravitational forces to push together gas and dust in certain areas, igniting vigorous new star formation, which appear as the spiral arms. The largest of the new stars are very bright, so the arms are rather spectacular when viewing a galaxy through the lens of a telescope. Spiral galaxies also often have groups of stars that are so far away from the core, they tend to follow much more chaotic orbits than the stars in the spiral arms. The farther out from the central pull of gravity, the less pull there is on a star to rotate around the core in an orderly orbit.

The Milky Way galaxy is one of these larger spiral galaxies, but with an added twist: it also has a bar formation of stars that looks like a condensed straight line of these stars running through the galaxy's center. If we could view the galaxy from outside of itself, we would see something like the image above. Not all spiral galaxies have this bar feature, and it was once thought that ours was without it as well.

The Milky Way formed more than 13 billion years ago in the middle of a cluster of galaxies called The Local Group. Two thirds of the way through the galaxy's life, our solar system was born from a collapsing cloud of gas and dust. Ever since then, the Milky Way has carried the solar system in orbit around the center, called the galactic core. The galaxy continues to produce stars; currently the rate of star production is at about seven solar masses per year, which means that seven times the mass of the Sun is produced.

Elliptical Galaxies – The Milky Way's Future

While spiral galaxies have spectacular structures, elliptical galaxies have structures that are rather unexciting to the eye. Ellipticals have no distinct pattern and are nearly featureless. They range in shape from almost spherical to somewhat flat. Their range in size is more diverse than other types of galaxies; from puny little ones with just a few million stars, to truly massive ones with up to several times the star count of the Milky Way, numbering into the tens of trillions.

Elliptical galaxies currently make up about 10-15% of all galaxies. Elliptical galaxies were not a dominant feature of the early Universe – their formation came later. Many of them are the result of dozens of smaller galaxies merging, probably including a few spirals. As the Universe ages, the percentage of elliptical galaxies will steadily increase. Mergers have occurred several times with the Milky Way already, when dwarf galaxies got trapped in its gravitational pull. Our galaxy has a nemesis on its doorstep, though. Andromeda is one of the few galaxies not moving away from our own. In fact, it's fast approaching on a direct collision course at about 402,000 kilometers per hour.

Fortunately, even at this frantic speed, it will still take about four billion years for it to collide with the Milky Way. After another few billion years, these two spiral galaxies will merge to form one massive elliptical galaxy.

Irregular and S0 Galaxies – The Unwanted Offspring

When the Milky Way collides with Andromeda, both our galaxies will undergo extreme changes that will tear stars from their galactic orbits, possibly flinging them far out into space for a time before gravity takes hold and pulls them back. These stars may keep going if gravity loses the tug of war. As you can imagine, changing from two spiral galaxies to one elliptical is going to be a chaotic process. Interestingly, merging galaxies almost never cause individual stars to collide with each other, owing to the vast distances involved. Stargazers on worlds in such an event may never know their home galaxy had merged with another.

Two other less common galaxy types may form as a by-product of the merger: S0 and irregulars. S0 is an intermediate type; its shape is not quite like either an elliptical or spiral, but a contorted version of both of them. S0s contain little gas and dust, have few stars, and are smaller and less bright than most other galaxies. Eventually an S0's shape will become more elliptical or spiral as the stars find a gravitational balance with each other. The merger between Andromeda and our galaxy may result in S0 galaxies forming around the newly merged main galaxy.

Then there are irregular galaxies, which, as the name suggests, have little to no consistent structure. Irregulars have bright regions full of hydrogen that are in the process of forming new stars, sticking out from the rest of the galaxy like a lighthouse on a stormy coastline. Eventually, irregulars will also gain a more defined elliptical or spiral shape.

Stars – life's parents

Carl Sagan once said, "We are made of star stuff."

This statement is quite literally true. The Universe is composed of roughly 75% hydrogen and 25% helium – everything else is in a sliver of a percent. Stars fuse hundreds of millions of tons of hydrogen per second into helium. As a star ages and uses up its hydrogen fuel, the core of the star contracts in order to maintain the fusion process. The more condensed and now hotter core of the star allows helium to fuse into heavier elements, including carbon and the other elements of which our bodies are composed. When we detect a star's composition, we are trying to determine its "metallicity" (astronomers call all elements heavier than hydrogen and helium "metals"). When the Milky Way first formed, there were no elements heavier than hydrogen and helium. As supermassive stars exploded, they released heavier elements generated in their cores, seeding the galaxy with them. Future generations of stars contained more and more of these heavier elements. Late generation stars with a high metallicity have a greater chance of hosting planets, as planets themselves require metals in order to form. Stars of a wide variety of sizes and colors dot the Universe, including extremely massive ones that are larger than our entire solar system.

These exceedingly immense stars will only last a few million years, which is far too short a time for life to evolve on an orbiting planet, let alone for life on any planet to build a civilization. Many of the first stars in the Universe were thought to be these massive giants, owing to the higher density of gas clouds during galactic birth. Our Sun is a typical G-class type star that converts hydrogen into helium, along with a few heavier trace elements. For stars the size of our Sun or smaller, carbon is about the heaviest element that is ever produced. For larger stars, especially the supermassive ones on the order of eight times or more of the Sun's mass, like Betelgeuse, it's a wildly different story. Elements as heavy as iron can fuse, at least for a brief period of time. Iron literally saps energy from the star's fusion process and never releases it.

This prevents the star's energy from pushing outward, allowing the constant inward pull of gravity to overtake. With less energy to continue pushing outward, the star collapses in on itself. As the star begins to shrink, each element separates from the others and settles into a shell shape around the core. First there is hydrogen fusing to form helium, and then lithium, beryllium, boron, carbon, and so on, in layers like that of an onion. The collapse process lasts just a few minutes before the star becomes unstable and explodes as a supernova in one of the most energetic events since the Big Bang itself. The explosion is so powerful, that in that instant elements fuse to other elements to create heavier ones. Iron fuses into cobalt, nickel, and continues all the way down the periodic table to about element 98, Californium. After that, even the might of a supernova is not enough to create heavier elements – those can only be produced in a lab. As the supernova explosion progresses, the dying star's outer layers and much of its core are torn apart and sent hurtling into deep space, peppering the galaxy with the gold on your necklace and the copper in your electronics.

The birth of a planetary system

In order for comets, asteroids, planets, and every other object in a planetary system to exist, heavier elements than helium are required. Since the early Universe was only composed of hydrogen and helium (and traces of lithium), the very first generation of stars had no rocky planets around them. I'll call these naked stars. Naked stars are still being born in the Milky Way today, but at a far lower rate than during the birth of the galaxy. This is mainly because most of the gas and dust clouds available to form new stars have already been seeded with heavier elements. As the remnants of stellar explosions like supernovae mix with nearby gas clouds, it will cause the clouds to start collapsing, eventually forming a new planetary system. These new stars will now contain heavier elements, thanks to the neighboring supernovae. As more supernovae occur, additional heavier elements will pepper nearby gas clouds. At least two other generations of more massive stars had to die before a star like our Sun could be born. It took hundreds of millions of years before metal-rich stars and planetary systems like we see today with our own solar system could exist.

Star classifications – life's sweet spot

Like galaxies, stars are classified based on their properties, including size, temperature (and thus color), and how they use up their nuclear fuel. About 90% of all stars you see in the night sky are part of what's called the main sequence. Stars in other sequences are either older supermassive stars like red giants, small white dwarfs that have already reached the end of their lives, or neutron stars that are superdense remnants of stars that blew up in a massive supernova. We want to focus on the stars that are of a stable age with plenty of energy left in them, able to support a planet with life and civilization.

The main sequence stars are young adults to older adults – these stars are in the prime of life. Designations of stars in the main sequence range from O, B, A, F, G, K, and M. The Os are the largest and hottest stars, with the comparatively tiny M-dwarfs being the coolest. Our Sun is squarely within the G-type. It's still a large star compared to tiny M-dwarfs (which are part of the M-type subcategory of stars), but it doesn't come close to the truly gigantic O-type, which can be as wide as our solar system!

O-, B-, and A-Class Stars – Hot and Heavy

The O-, B-, and A-type stars are lumped together here because the chances that any orbiting planets will be habitable for long enough to evolve complex intelligent creatures is close to zero for all three of these types. O-type is hotter than 30,000 Kelvin, while B-type is between 10,000-30,000K. Type-A broils between 7,500-10,000K. The hotter the star is, the bluer it is. These three classes also have the greatest mass and brightness of all main sequence stars. A-types appear white to bluish-white to the naked eye and have a mass about 1.5 to 3 times that of the Sun. Examples include the famous Vega star, mentioned in the movie Contact. While all stars have zones where simple life forms could theoretically survive, and probably have the planets on which life can form, A-type stars live less than a billion years before blowing up. A billion years is only enough time for a planet's crust to cool and form a stable layer of liquid water on its surface. Thus A-type stars' lifespans are too short for any planet orbiting them to become habitable. The chances of habitability are even worse around the hotter B and O type stars. Fortunately for the chances of life amongst the totality of stars out there, O-, B- and A-type comprise only about 1% of all stars.

While these giants are not good candidates for the development of complex life around them, they are important for generating supernovae. As explained above, these cataclysmic events are the "heavy lifters" that create and disperse important elements needed for the chemistry of life.

F-Type Stars – On the Edge

F-type stars are classified as yellow-white dwarfs; they are hotterthan our Sun, but not as hot as O-, B-, or A-types. F-types are on theedge of being able to support life. Their lifetimes are short – only about2-4 billion years, depending on their mass. It is not entirely out of thequestion that life could evolve on a planet orbiting an F-type star, butthat life would need to evolve quickly, before the star changes too muchto destabilize an orbiting planet's biosphere.

On Earth, it took at least two billion years for life to get to just themulticellular stage. Complex animal life, including the development ofthe central nervous system, took yet another couple of billion years. Itmight be that as soon as complex life is forming on planets orbiting Ftypestars, the stars are already entering a late stage of development,and heating up so quickly that they end up snuffing out any budding lifeon their planets.

G- and K-Type Stars – The Sweet Spot

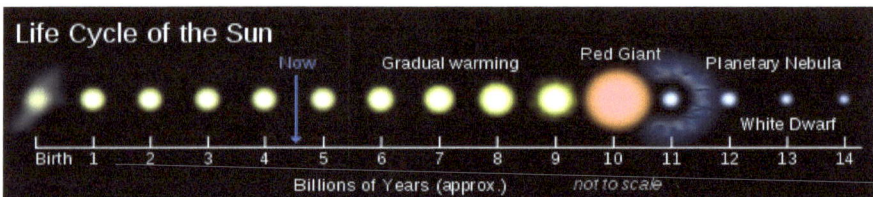

We now get to the most promising life-supporting candidates withG- and K-types. Both are smaller, cooler, relatively dimmer, and longerlivedthan the hotter types of stars, giving life a longer period of time toevolve into something interesting. G- and K-type comprise only about10% of all stars in the Milky Way though. They nonetheless still numberin the billions.While our Sun is commonly known as a yellow dwarf when viewedfrom Earth, this is a misnomer, as it is actually white when viewed fromthe undistorted perspective of space. The appearance of it being yellowto our eyes is due to the atmosphere distorting the incoming light. Evenwithin the whole range of G-types, there's only a hint of yellow forslightly less massive stars than our own Sun.The lifetime of a star similar to the Sun is going to be about tenbillion years. Although this is much longer than the relatively brief twobillion years of F-type stars, it still doesn't really provide a lot of wiggleroom for life to evolve, at least when compared to the coolest of stars.Regardless, ten billion is obviously still enough time for a civilization toform and have a good chance to explore the Universe, as here we are!

The size and lifetime of a star also has a strong correlation to itsbrightness. Because G- and K-type stars are at the upper sizethreshold to support complex life, any inhabited planets in orbit willlikely have skies not significantly brighter than our own. The averagehabitable world may actually have considerably dimmer skies, as wetake into account K- and M-type stars.K-type stars are just a bit cooler and dimmer than our own Sun.They have one of the biggest benefits for life to evolve – they lastlonger than G-types. K-types will last at least ten to twenty billion years.It's plenty of time for life to evolve, be destroyed and evolve again a fewtimes over – definitely enough time to develop a civilization that canreach for the stars.As a bonus, the habitability zone of K-types would not shift asrapidly as the hotter stars, thanks to how their energy output increasesmuch more slowly as they age much more slowly.K-type stars are my favorite candidates for life because of the aboveattributes, not to mention you wouldn't need as strong of a pair ofsunglasses. We know that life can exist around G-type stars like ourSun, and K-type are nearly the same. Their planets will have to orbit alittle bit closer for warmth, but that won't be a problem for at least thelarger of the K-type stars.

M-Type Stars – An Enigma

As we know from our understanding of evolution, life requires veryspecific conditions for civilization-building creatures to evolve. This putsinto question life around M-type stars.M-type stars are different from all other star types in just aboutevery way, both with positive and negative consequences for life. M-typesmainly include, among others, M-dwarfs, also called Red Dwarfs,which are the smallest of stars in the main sequence. M-types can alsobe much rarer giants, if they are stars at the end of their life. Focusingon M-dwarfs for the purpose of hosting habitable planets, they have twobig advantages: their longevity and their abundance.The ability of M-dwarf stars to consume their fuel at a slower rate isthe main key to their longevity, some of which can last for trillions ofyears. In other star types, the convection process is limited to either thecore or the outer layers, never to the entire star. In an M-dwarf star, thehydrogen mixes throughout its entire structure. As such, the star canlast significantly longer before running out of fuel.In addition to the longevity of M-dwarf stars, the fact that they are byfar the most numerous makes them exciting candidates for hosting lifeon orbiting planets. As many as 80% of stars in the galaxy are M-dwarfs.Even if life were a fraction as abundant around M-dwarfs asother star types, we should still expect to find several times more lifebearingplanets around these stars than all other star types combined.Because of an M-dwarf's tiny size, any planets must orbit very closeto keep warm and to retain liquid water on their surface. Theunfortunate side effect of the planet being so close, though, is itsatmosphere is touched by the star's flare activity.M-dwarfs are extremely unpredictable in the amount of deadlyradiation they produce, at least early on in their lifetime.

In a sun-likestar, flares develop from convection in the outer layers. With an Mdwarfstar, the entire star remains in this twisting and convective dance,so the magnetic field lines become much more contorted, capable ofthrowing out frequent and extremely powerful flares that can test aplanet's magnetic field in protecting life on the surface.

The stellar graveyard

Stars eventually die. They last from just a few million years, like theO-type giants, up to trillions of years for the flaring M-dwarfs. As theyage, the flaring activity calms down, at which point the stars enter amidlife period of relative stability. Our Sun is currently in this midlifestage, at 4.5 billion years of its 10 billion-year lifespan.The hydrogen the star relies upon for fuel has to run out at somepoint. Depending on the original mass of the star and how it convectsthe fuel throughout the star's layers, the fate of stars varies. Theirfusion process will eventually stop and for the non-exploding largerstars, what's leftover is a white dwarf star. A white dwarf is a small corethat no longer produces fusion, but still shines in the night sky. It willcontinue to radiate heat for billions of years, to finally rest as a deadblack dwarf – a former white dwarf star that no longer emits anysignificant heat or light.For our Sun, and stars of a similar type, as they get hotter, itshydrogen fuel depletes, and its core eventually compresses as it startsto fuse helium into carbon. This will heat up the outer layers, causingthe star to expand to many times what we see it today. This expansionwill come in fits and starts, progressing until its atmosphere cools andturns an orange-red.

At its largest, the Sun's radius will be about as wide as Earth's orbit. The Sun will either engulf the planet, or cause it to migrate to a more distant orbit. The process of expansion will begin long after the Sun boils off Earth's oceans, leaving our planet a dry, lifeless tinder. While complex life will be long extinct, it is possible that the Sun will have shed enough mass during this time period, causing the gravitational pull to weaken, which will in turn allow planets to migrate further out. Our planet may end up nearly where Mars currently orbits. If Earthlings have colonized Mars before Earth becomes uninhabitable, then we might be able to terraform Earth and bring it back to life again. For the largest of stars, they will form either ultra-dense neutron stars or a black hole. Neutron stars are fantastically dense, and due to conservation of momentum, can rotate hundreds of times a second. The gravity well is so strong that the atoms inside are crushed to the point that the electrons and protons fuse into neutrons, thus the origin of the star's name. A single teaspoon of neutron star material would weigh up to a couple billion tons, or half the weight of Mount Everest!

Habitable zones

As we know, not all planets are habitable – just look at our own solar system as an example of the limitations around an ordinary and relatively stable star: only 1 out of 8(+) planets can host life. Mercury is simply too close to the Sun, Venus just missed the mark of habitability, Mars had a chance for a while but it is too small to retain an atmosphere, and the rest of the planets are distant giants with a crushing atmospheric pressure hundreds of times that of Earth. Sometimes called the Goldilocks Zone, the habitable zone is the area around a star, or possibly a gas giant planet, where life has the chance of forming. The boundary begins and ends where liquid water can exist on a surface with atmospheric pressure, i.e. a planet or moon that has a thick enough atmosphere to sustain liquid water. Too close to the star and water would boil off, while too far away and it would freeze. The hotter the star, the wider and further out its habitable zone extends. One of the redeeming traits of F-type stars is their zone is much wider than our Sun's own zone, and yet the star isn't too hot to overly restrict life's chances to appear. The zone of F-type stars is also farther from the stars' dangerous solar flares. A wider zone provides a greater chance for a planet to have liquid water on its surface, and thus be habitable.

In a billion years, Earth will no longer have water because of the Sun's increasing luminosity. Every star heats up over its main sequence lifetime, and at ever faster rates for these hotter stars. For M-dwarfs, the habitable zone is going to be extremely narrow, and very close to the star. Any habitable planets need to practically hug the star, perhaps as close as .10 au, or about 10% the distance from which Earth orbits the Sun. The astronomical unit (au) is the distance from the Sun to Earth – 150 million kilometers, or 93 million miles. We use this measurement to judge distances between bodies in our solar system, including other systems.

Venus and Mars – A Unique Family

Defining the boundaries of a habitable zone has been a heated debate recently, now that we've been discovering so many unique types of planetary systems. A lot has also been learned about how carbon dioxide, methane, and even water vapor act as powerful greenhouse gases that help to regulate a planet's atmospheric temperature. A planet's ability to retain heat with greenhouse gases and a thick atmosphere, will either expand or constrain the habitable zone of other systems. If Earth and Mars were swapped, it is possible that Earth in Mars' position would still be a wet and warm planet because its larger size allows it to retain a thicker atmosphere, trapping in more heat. For Mars, though, it's uncertain if it would be habitable in any position within the solar system. The smaller the planet, the more quickly it loses its ability to retain heat, generate plate tectonics, and hold on to its atmosphere. If the atmosphere is lost, so will the surface water that is essential for life to evolve. Mars also lacks a magnetic field to counteract the solar wind that would also strip away the atmosphere.

Physically, Venus is comparable to Earth in many ways, yet the planet is devoid of life. When the Sun was 30% cooler at the formation of the solar system, Venus would have been well within the inner edge of the habitable zone, much like where Earth is today within the zone.Instead, Venus is now outside this boundary, too far inward toward theSun. Our planet has been completely frozen over many times in itspast, so a Venus in Mars' orbit might look very much like Earth today, orone of its past snowball periods.Plate tectonics are critical for a planet to be able to recycle itsatmosphere and keep carbon dioxide, methane and other greenhousegases at balanced levels. Plate tectonics provides a natural thermostatthat could allow a planet to survive outside the normal habitable zoneboundaries. Scientists think that on planets at the lower end of themass scale, like Earth, water is a required lubricant to keep the platesmoving. Without water, the plates would literally get stuck and the platetectonic process could shut down. Simulations suggest that a largerplanet than Earth may be able to retain active plate tectonics,regardless of its water supply.The shutdown of plate tectonics is probably what happened toVenus after its surface water evaporated. The poor planet was simplytoo close to the Sun and any oceans it had boiled away. In addition, theplanet's extremely slow rotation (likely due to an ancient collision with aprotoplanet), and different internal cooling properties, may haveprematurely ended its ability to generate a global magnetic field, whichin turn helps in protecting the atmosphere and limits water fromescaping into space.Once Venus lost any surface water it had, the water vapor in theatmosphere started to act as a greenhouse gas. It rose to thestratosphere and split apart into hydrogen and oxygen from the intenseultraviolet sunlight. The hydrogen escaped into space, whereas theoxygen combined with carbon to form carbon dioxide. With platetectonics non-existent, the carbon dioxide would never be absorbedback into the mantle. The end result would be a buildup of carbondioxide in Venus's atmosphere to today's level of 96%.A constant loss of atmosphere is what is happening with Venus andMars as well. We can even see the atmosphere being blown into spacewith certain probes we've sent to these planets, such as NASA'sVeSpR spacecraft sent to Venus, and NASA's Maven probe sent toMars. With a non-existent magnetic field that would otherwise keep thesolar wind at bay, a strong electric field still exists and may be aiding inthe loss of atmosphere and its water. This is especially the case for thenow very dry Venus. The electric field pushes up water molecules highinto the atmosphere, where they are then carried off by the solar wind.As a consequence of Venus's failure to retain a life-supportingatmosphere, the planet today is one of the most inhospitable rockyworlds in the solar system. It has an iron-melting surface temperatureof 462 C (863 F), winds in the upper atmosphere as

41 | Guide to Our Galaxy

high as 400kilometers per hour, and a crushing atmospheric pressure 92 times thatof Earth. The pressure on the surface of Venus is equivalent to thepressure a kilometer below Earth's oceans. Life could very well havestarted on Venus billions of years ago, but, sadly, complex intelligentlife would never have had enough time to develop.

A galactic habitable zone

While there are zones around the Milky Way that tend to producestars with greater metallicity, and thus also planets, it's not as easy todefine as recognizing a planet is within its star's habitable zone. Still,we can calculate some estimates based on a galaxy's structure.Generally, the farther a star is from the center of the galaxy, the lesslikely it is to have been seeded with heavier elements. Far out stars inthe galactic halo are called Population II type stars. Our Sun is aPopulation I type star and is located well within the galaxy's main bulkof stars.The habitability of a planetary system within a galaxy is determinedby two primary factors though. The first is whether or not a star hasenough heavier elements so that planets can form. The second factoris whether or not a star is located in a region of the galaxy devoid ofdestructive phenomena, such as supernova and gamma ray bursts(GRBs). If both of these criteria are met, then we could say that aplanetary system is within the Galactic Habitable Zone (GHZ), at leastfor as long as it stays in the zone.

All stars in the galaxy are moving, sothis safe status for any system is likely temporary, including for our ownsolar system.Some star systems have more dangerous paths through the galaxythan others. Occasionally two stars will drift close to each other,perturbing their surroundings and threatening any habitable worlds withincoming asteroids and comets. Our own solar system has a cloud ofprimordial comets and asteroids far beyond the orbit of Pluto, called theOort Cloud. The material orbits over the course of thousands of years,rarely disturbing the inner planets. When it gets disturbed, though, itcan be devastating. At least one of Earth's great extinction events isthought to be caused by a star passing by the Oort Cloud and knockingcomets and smaller asteroids into the inner solar system.Another extinction event in Earth's history is thought to have beencaused by the solar system being too close to a supernova or a GRB.Fortunately, supernovae only occur in a galaxy about three times everyhundred years, and GRBs occur far less frequently than supernovae – only about three times every million years. The energy from a GRB alsoneeds to be directed towards a planet, as they do not spread out theirradiation in all directions as much as supernovae do. If a GRB everdoes hit Earth, it would take as little as a few seconds for it to destroyour protective ozone layer, without which nearly all higher forms of lifewould die. Millions of years would need to pass for life to recover andbecome as grand as we see it today.

Fortunately for us, our solar system resides comfortably betweentwo galactic arms as it revolves around the galaxy. Like the galacticcore, these arms are areas that form more massive stars which willeventually explode violently. Our solar system only rarely crosses thesehigh risk areas. Other systems follow much more chaotic or elongatedpaths. Some of them will remain in a galactic arm for millions of years,causing habitable worlds to be hit with multiple radiation events,perhaps resulting in their becoming sterile permanently.Sadly, the neat little arrangement our solar system has with theMilky Way is only temporary. Once Andromeda collides with our galaxyin about four billion years, every star system will be flung in differentdirections. Some will even be ejected out of the newly formed ellipticalgalaxy altogether. A galactic habitable zone may become ameaningless term in this chaotic scenario.

Alien planetary systems

In the search for distant astronomical objects like exoplanets(planets outside our solar system, also known as extrasolar planets),telescopes are a critical tool. The first generation of telescopes used forplanet hunting were built in places like Hawaii, on top of mountains,such as the Caltech Submillimeter Observatory (CSO), built in 1985.Since a telescope has to peer through Earth's atmosphere, it needs tobe as clear of as many observational distortions as possible, such aspollution, including light pollution. Humidity can also interfere withclarity, which is one of the key reasons that mountain tops makeexcellent sites for telescopes.Discovery of the first exoplanet, 51 Pegasi b, was made in 1994 byDr. Alexander Wolszczan, a Polish astronomer at Pennsylvania StateUniversity. He discovered what is called a "hot Jupiter" orbiting 51Pegasi, a star in the Pegasus constellation. Hot Jupiters are massiveplanets the size of Jupiter that orbit extremely close to their host star.

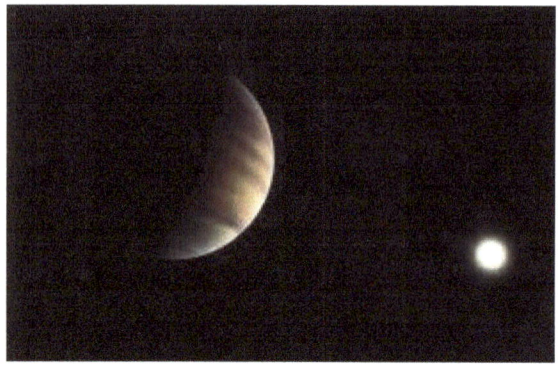

While the discovery of 51 Pegasi b was exciting news, the planet turned out to not be a candidate for supporting life. Not only was the planet extremely close to its host star, but the star itself was a pulsar. Pulsars are dead stars that previously exploded in a supernova; what remains is a super-dense core that spins rapidly, giving off radio pulses as it rotates.

Unfortunately, there is little hope of life around such a hostile environment. Even though life may not exist on 51 Pegasi b, the discovery of the hot Jupiter proved that other star systems contain planets. The discovery also proved that at least some of those planets can survive the death throes of their stars, or possibly even be created among the debris left over from the supernova aftermath. Just a year later, in 1995, the Swiss team of Michel Mayor and Didier Queloz discovered another planet, but this time around an ordinary star like our Sun. The planet was about the size of Jupiter, and orbits its star so closely that it makes one revolution in just over four Earth days! Its atmospheric temperature is hot enough to melt lead, making it impossible for any spacecraft to survive on the surface for more than a few seconds. Still, it was progress in finding a planet like our own.

The golden age of planet hunting

Planet hunting is multitudes more challenging than trying to find an actual needle in a haystack. Our ability to confirm planets varies with the size of the planet and how far out it orbits from its star. The smaller the planet, the harder it will be to detect, and the farther out the planet orbits, the less its effect on the parent star can be confirmed. As planet confirmation took off late last century, scientists, eager to detect the first exoplanet, worried that we may never find a planetary system like our own, regardless of the detection method. Before the first exoplanets were finally discovered, there was an expectation of how a system ought to be configured: the small rocky worlds would reside closer to the parent star, gas giants and icy worlds would inhabit the outer regions, and somewhere in the middle would lie an asteroid belt – a nice arrangement that provided lots of room for a planet (or many) within the habitable zone. Except that is not what has been discovered.

Discoveries so far suggest just about every configuration possible but one like our solar system. The very first planet discovered, 51 Pegasi b, orbits in just a few days around a dead star. Most systems also host super-Earths, which are planets that are just a bit larger than Earth. We have also discovered planets with highly elliptical orbits that make possible any other planets in the system doubtful; as a planet's orbit becomes less circular and more elliptical, the planet has a greater chance of crossing paths with another planet. Even binary stars (two stars closely orbiting each other) have been shown to host planets.

Detection techniques

One of the first planetary detection techniques used, the radial velocity method, detects the tug of a planet on its parent star. The radial velocity method requires the planet to be extremely close to its star, because we're relying on very tiny readings of how the star is being affected by the planet gravitationally, not through observing the planet itself. This method is very accurate in estimating a planet's size and distance from its star, but it is not able to tell us much about a planet's composition, including what gases make up its atmosphere. The most successful method to date has been the transit photometry method. As a planet passes in front of its star, the overall brightness of the star dims by a tiny fraction of a percent. The larger the planet that passes in front of its parent star, and the dimmer the star, the easier it is to detect a dip in overall brightness. The Kepler Space Telescope was built specifically for this technique. With it, we have been able to identify thousands of planets with thousands more yet to be confirmed. Confirmation takes several revolutions of the planet to ensure no false positives. The transit method has an advantage over the radial velocity method in that it can detect planets further from the parent star, out where the habitable zone lies. Other advantages of the transit method are that planets as small as Mars can be detected.

The transit method is our current best method for discovering a habitable world similar to Earth. There are a few disadvantages with the technique. At most, only 10% of all stars and their planets will be aligned in such a way that, from our vantage point, we can see any planets pass in front of the star. Kepler has a field of view of more than 145,000 stars, yet it has only confirmed about 2,300 exoplanets so far. Another disadvantage is that confirming the existence of planets requires at least three detected orbits. If astronomers are trying to confirm a planet around a star like our Sun at the distance Earth orbits, then it will take a year before a planet will pass in front of the star from our point of view. To detect a planet as distant as Saturn in this way, it would take 29 years! Confirming a Saturn world would thus take nearly a century in Earth years.

Other more advanced techniques include direct imaging of a planet by blocking out the light of the parent star. This approach is one of the most promising techniques because of its ability to image planets directly, regardless of how a planetary system is aligned with our point of view. The technique uses a star shade (thin film of material floating in space) to block out the light of a star, allowing us to see any orbiting planets with a camera positioned behind the star shade. Another method uses a quirk of physics called gravitational microlensing to peer through a planetary system to image a planet directly.

Gravitational microlensing works much like the lens of eyeglasses, except it's a star that's bending the light. The light from a star travels towards another large object, say another star, which causes the light to bend around the object in such a way that focuses that light more than it would be traveling in a straight line. That focusing of the light allows us to more closely evaluate the star's properties, and because planets are close in orbit, find the light reflected off the planet as well. Whichever method is used, planet detection is a very difficult and sensitive process. It's truly amazing what planet hunters can tease out of the data despite the accompanying interference. For example, to find a typical Earth-sized planet around a sun-like star using any of these methods would be akin to having a firefly in front of a spotlight in San Francisco, while you are in New York and using your unaided eyes to try and see the firefly.

A new generation of telescopes

Most telescopes are capable of numerous types of observations, including planetary research, but also Earth based science like meteorology.5 Scientists bid for their use, especially on the larger and more powerful telescopes. Up until 2009, we had no telescopes exclusive to planet hunting, and certainly none launched into space. Investors didn't want to incur the expense of launching a telescope just on the seemingly slim chance that a few exoplanets would be found. There were means of detecting planets with existing telescopes, so we first used those to make initial discoveries that then justified further equipment. Once it was clear that exoplanets existed in abundance, the Kepler Space Telescope was launched to discover more. The great news out of the data collected so far is that just about every star seems to host at least one planet, and probably a handful more. This includes binary star systems, as well as smaller M-dwarfs. As detection techniques are refined, we're discovering ever smaller planets than even Mars and, most importantly, at orbital distances where liquid water may flow on the surface. There are some very exciting telescopes being built, due to start operations this decade, many of which are dedicated to planet hunting. The most sensitive new telescopes are those that will be launched into space. Two new ones will replace existing telescopes with far more powerful instruments: the Transiting Exoplanet Survey (TESS), and the  James Webb Space Telescope (JWST). There are also quite a few being built on the ground, such as the W. M. Keck in Hawaii and the Atacama Large Millimeter/submillimeter Array in Chile.

TESS was launched in 2017 and is the successor of Kepler. It also uses the transit photometry method of detection. As mentioned earlier, this method probes stars for planets that pass in front of their light. The benefit of this method is that we can understand in detail the planet's overall size, mass, water content, atmospheric density, composition, and even any industrial pollutants in the atmosphere. The JWST launched in 2018, the successor to the Hubble Space Telescope. Much like Hubble, JWST is tasked with other research priorities, not just planet hunting. The telescope is the most advanced we've ever sent into space, and also the costliest, at more than 8 billion USD. At several times Hubble's size, it needed to be folded up and then unfurled in space. As you can imagine, the process of unfurling is risky, so rigorous testing was done in order to get it right. If the telescope malfunctions once in space, it will be so far from Earth, at an incredible 1.5 million kilometers that it will be practically unrepairable without costly missions that take years to complete. The JWST will be observing objects in the infrared wavelength of the electromagnetic spectrum. It will therefore need to be far away from our Sun, otherwise it will pick up the infrared portion of the Sun's energy, drowning out any signal of a distant planet.

The telescope itself will also have to be extremely cold, otherwise its own electronic heat will distort the image; in fact, it will be operating at –225 °C, or 373 °F below zero! The best location for the JWST is what's called a Lagrange point. These are points in space between two large bodies, like the Sun and Earth, where the gravitational pull between the bodies is balanced. Place an object there and it will stay there, instead of being pulled more strongly one way or another. There are five such gravitationally stable points around any two large bodies. Only one Lagrange point, L2, will be a suitable location for JWST. For Earth and JWST, L2 is located on the far side of the planet from the Sun, and beyond the moon's orbit.

What will we find?

Our sciences have come a long way since Galileo discovered in 1610 the first moons orbiting another planet, around Jupiter. We understand a lot about the basic makeup of the solar system, the Milky Way galaxy, and the farthest reaches of the known Universe. There seems to be a lot of commonality when we start categorizing things like galaxies and stars, but once exoplanets were discovered, astronomers quickly realized how much variety the Universe has yet to reveal to us. We will explore the possibilities in the next chapter.

CHAPTER 4: THE BOUNDARIES OF HABITABILITY

Stars, planets, and life itself arise from a Universe governed by universal laws of nature. Life on Earth has evolved in accordance with these laws, and alien life elsewhere will do so as well. That alien life may have a stockier body due to its planet's stronger gravity, or larger, yellow-tinted eyes due to a slightly different atmosphere, but there's a reasonable chance it will have a body and eyes. Even different civilizations' inventions will have common properties, owing to the laws of physics, economic constraints, and other universal conditions.

In this chapter we will explore a variety of worlds that may have a chance at being habitable. A continued reference point is our own solar system we've come to know and love. Starting with the inner rocky planets, including Earth, they share many properties. They all orbit in a nearly circular path around the Sun, each has an atmosphere, and a solid surface that one could set foot upon, a day and night cycle, and seasonal weather patterns. They are also all rich in organic compounds, including the chemicals that emerge from active (or previously active) volcanic systems. Life is not found on any planet except Earth, though. These sterile worlds seem to be the norm rather than the exception, even with so many shared properties.

As we talked about in the previous chapter, there are two primary considerations for planets to be potentially habitable (moons will be investigated later): a planet's distance from the parent star and the planet's size. These parameters are important in every system, even binary or trinary star systems. The orbital distance of a planet has to allow for liquid water on the planet's surface. Too close to the star and water will evaporate into space. Too distant – beyond what's called the Snow Line – and water will freeze. Regarding size, most smaller planets within a star's habitable zone, like Mars, will not be able to support a thick enough atmosphere or protective magnetic field, and gas giants have atmospheres that lack many of the minerals thought needed for life to evolve.

Encouragingly, of all the stars we've looked at so far that have planets in orbit, it seems the average number of earth-sized planets found in the habitable zone is at least one per star. The excitement can be tempered with the fact that Mars, and nearly Venus, also reside within the Sun's habitable zone, and one wouldn't want to book a vacation to either of them anytime soon.

Without closer inspection of planets, what we think we know about them can be invalidated with the next discovery. For example, the planetoid Pluto was thought to be a completely sterile world devoid of even the thinnest of an atmosphere, with no geological activity of any kind – even less life-friendly than our own moon. What we in fact found though was that Pluto has an active geological system with ice mountains thousands of feet high, cliffs and troughs stretching hundreds of kilometers, and a thin atmosphere that snows nitrogen in regular seasonal patterns. Even with these active systems, Pluto unfortunately still seems to be a sterile world. The more we learn about the worlds in our solar system, the more we're both surprised at their variety, and also disappointed at their revealing harsh constraints on where life can appear. This surprise and disappointment duality is likely to repeat itself as we continue to explore other planetary systems.

Introducing super-Earths

In between gas giants like Jupiter and the tiny dwarfs smaller than Mars, there is a category of planets larger than Earth, called super-Earths. Super-Earths include both rocky worlds, as well as mini gas giants that are similar to Neptune and Uranus, but smaller. So far, the possibility of life on these earthly giants is looking both promising and a bit uncertain. The larger super-Earths are believed to have extremely thick atmospheres of hydrogen and helium, near the density of Venus's atmosphere. The smaller super-Earths, however, up to about two times the mass of Earth, are believed to be good candidates for life, despite still having somewhat of a thick atmosphere. The smaller super-Earths are much more likely to have a solid surface as well. Scientists are close to being able to detect the composition of the atmospheres of these worlds, and thereby gain great insight into their habitability.

Gravity of super-Earths

Once a planet's mass and volume is calculated, we can easily estimate its gravity. Let's assume we discover a planet tomorrow that has the same density as Earth, but with a radius 25% greater. This would put it well within the super-Earth category. It would have about three times Earth's gravity. That sounds crushingly uncomfortable, and it would be to us, but not necessarily for any life that evolves to adapt to that pressure. Fighter pilots can handle forces greater than 3G (three times Earth's gravitational pull). Human physiology only starts to have problems upon reaching 5Gs for more than a few seconds. At about 8-10G, we risk passing out; the heart becomes unable to pump blood to the brain.

There are surprisingly few obvious physical limitations that would prevent a civilization from developing on a stronger gravity world. The most significant issue would be the civilization's ability to send objects into space. Even a modest increase in gravity would noticeably increase the fuel required to launch a rocket into orbit. On Earth, it already costs thousands of dollars to send even a few pounds into orbit, and over 90% of the mass of a rocket is in the propellant. Other than the interest to explore outer space, a civilization would only need to compensate for a higher gravity environment, such as using thicker steel for skyscrapers, and lighter composite materials for airplanes would probably be researched sooner than they were on Earth.

Geology of super-Earths

The geology of a planet is critical to maintaining a stable environment in which life can thrive. Super-Earths may be even more geologically active than Earth, with vigorous plate tectonics to recycle their atmospheres and keep them cool enough for life to thrive. One might think that a vigorous shifting of the plates would make these planets extremely unstable. The case may be instead that because they can vent their interior heat more frequently, globally impactful events like volcanoes and earthquakes would occur less frequently.

Atmospheres of super-Earths

How exoplanets' atmospheres would (or do) affect any life on them is perhaps one of the most complicated and least understood aspects, even more so than what lies beneath their surface. Until super-Earths' atmospheres can be studied up-close and in detail, models are all we have to reveal the likely limits on life, and those models are currently showing a wide range of possibilities.

Most models of super-Earth atmospheres suggest that the smaller super-Earths have a thin nitrogen-covered envelope, while the larger ones are more likely to remain shrouded in a thick blanket of hydrogen. Some models suggest that a planet 1.5 to 2 times the mass of Earth would have so much extra hydrogen and helium that even with the star's ultraviolet radiation stripping away the atmosphere atom by atom, wouldn't be enough to remove all of the lighter gases. Our planet's own atmosphere is composed of 78% nitrogen, 21% oxygen, and traces of argon and other gases. It contains no hydrogen at all. As with a stronger surface gravity, an atmosphere denser than Earth's doesn't preclude life, but we are still unsure how a civilization may survive on such a world. (Additionally, the atmosphere is the final threshold to space, and a dense atmosphere makes it a challenge to develop further as a spacefaring civilization.)

The surface temperature of a world is an important factor in habitability. Cloud albedo (when clouds reflect sunlight back into space) regulates a planet's atmospheric temperature, so it plays a significant role in whether or not a planet is habitable. If we detect a planet with a high albedo specific to cloud formation (not a world covered in ice), then we'll know it's a wet world, and probably on the warm side. A greenhouse world would eventually result in the total loss of water on the surface of the planet.

Another factor of habitability of a super-Earth is its atmospheric windspeed. The faster a planet rotates, the stronger the average surface wind speed will tend to be. For example, Uranus, a gas giant much larger than super-Earths, has a rotation rate of about 17 hours per day and winds of hundreds of kilometers per hour. All else being equal, super-Earth wind speeds should fortunately be just a bit more than we experience on Earth, so life may still be able to thrive on these worlds.

Super-Earths are everywhere!

Super-Earths are extremely common in the galaxy; they are found around nearly every star that we detect orbiting planets. Interestingly, our solar system seems to be the odd one out in that it doesn't contain a super-Earth, while most other systems have at least one. Since Earth is thought to be on the smaller end of planets capable of supporting life, the law of averages suggests that most alien civilizations reside on these larger super-Earths, if these planets are found to actually be habitable at all.

Pushing the boundaries of earth-like planets

While there is a significant range in star sizes, the vast majority sit in a single category. As we talked about in the previous chapter, the largest stars (types O, B, and A) will burn themselves out before complex life has a chance to evolve on their orbiting planets. They account for about 1% of all stars. F-, G- and K-types add up to about 10% of all stars. Another 10% are dead stars and other oddities. We are left with nearly 80% of all stars that we are not quite sure can support habitable worlds at all.

This entire 80% rests with M-dwarf stars. They are so plentiful in the Universe that their sheer number alone demands that astronomers rigorously investigate the potential for habitability. There are two reasons they populate every corner of the galaxy: they last a long time and are produced in stellar clouds like weeds in a garden. Much like any objects found on Earth, it's typically easier for nature to produce the smaller variety. The tiniest of M-dwarfs are not all that much larger than the planet Jupiter.

So far M-dwarfs are proving very promising for finding planets around them. The question remains as to whether or not those worlds are habitable. Habitability is dependent more on the size of a planet than on any other factor. The size of all detected planets around M-Dwarfs so far range from gas giants like Jupiter all the way to Mars sized bodies (and undoubtedly there are smaller exoplanets, we just haven't detected these worlds yet), so size won't be a problem. We have reason to get excited that at least some of those worlds may be habitable.

An M-Dwarf's Younger Years

For the first couple of billion years or so of an M-dwarf's life, the star will go through the same active flaring our Sun did in its youth, except at an increased rate with even more powerful episodes. The intense output of radiation from these flares may severely disrupt the chances for life to thrive on an orbiting world, both because of the radiation itself, and because the star produces more sunspots during this active time. Sunspots are areas on the surface of a star where the star's magnetic field becomes twisted, producing intense energy that is ready to be unleashed at any moment in the form of powerful flares. The surface of a sunspot is actually much cooler than the material beneath. On stars like our Sun, sunspots can increase its light and heat output –possibly for months at a time. These sunspots can be a significantcause of climate shifts, such as Earth's Maunder Minimum event thatlasted from approximately 1635 to 1720 A.D.3 During this period, as thenumber of sunspots plummeted, so did the planet's averagetemperature.

Sunspots on M-dwarf stars function differently; they can be soenormous that their cooler surface areas cause the opposite effect, inthat they lower the light and heat output of the star. This reduction inlight and heat can be devastating to any life on orbiting planets byorders of magnitude of what Earth experienced during the MaunderMinimum event.

What would it be like if an M-Dwarf star's sunspots blocked out 10,20, or even 30% of the incoming light for months at a time, droppingtemperatures on an orbiting planet not by just a few degrees, butpossibly by hundreds of degrees? By comparison, if this dimminghappened to Earth, many plants and animals would suffer severefrostbite and eventually – or suddenly – die. The plants and animalsthat hibernate when winter arrives would need to hibernate for longerperiods. If life is possible on such a world, that life may produceamazing new features and abilities to adapt to extreme andunpredictable changes in temperature.

At its most active periods, our Sun spits off as many as twenty flaresa day. Young M-dwarfs may flare hundreds of times a day and emitgiant flares that temporarily double the star's brightness. Solar flaresare bad news for life, especially when the flares are hundreds of timesmore powerful than those our own planet has ever experienced. Flarescan overpower a planet's magnetic field, especially on a smaller worldsimilar in size to Mars where the field is probably going to be weaker.The atmosphere of a small planet could be stripped off by the solarwind. Life may never have the chance to even get started.

A larger planet's atmosphere though could survive an attack by theparent star's flaring activity. A super-Earth with a radius 1.2 to 2 timesthat of our planet may have an atmosphere that is dense enough for athin but life-sustaining layer of gas to remain, post-flare. An M-dwarf'smore active early years with its constant solar flaring may help clearaway some of the lighter gases. The remains of the atmosphere billionsof years later may then be earth-like. Then again, the parent star mayleave the planet a dry tinder if the flaring activity goes on for too long.

Tidally locked worlds

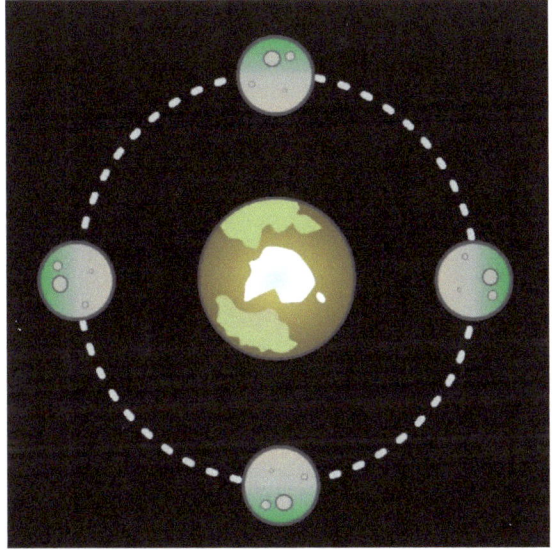

If you look up at our moon on a clear night, and do this repeatedly over the course of several evenings, you will notice that the same side of the moon faces Earth each night. How could this be if everything in the solar system is rotating? Shouldn't the moon be rotating as well? Yes, and it does, but we don't notice it because the rotation is exactly in sync with our planet's own rotation. Like two ballerinas staring into each other's eyes, they are barely aware that it's not the room that's spinning, but they themselves.

The moon's rotation has slowed Earth's rotational rate considerably since the moon first formed. At its formation, our planet used to spin so fast, a day was less than twelve hours long! Over time, the gravitational pull between Earth and the moon, as well as other bodies in the solar system, caused the rotation to slow. While an Earth day (one rotation) has doubled in time, the change has been even more pronounced for the moon because of the difference in size – Earth's gravity influences the moon more strongly. The moon now rotates so slowly, it is in what is called a tidally locked position, rotating only once for every revolution around our planet. This 1:1 orbital resonance is why the moon will always face the planet in the same way.

Any planet that orbits its star at half or less the distance that Earth orbits the Sun will also become tidally locked. The day and night cycle we find natural on Earth would not exist on these worlds. The star would forever shine in the same spot on the near side of the planet with the far side in perpetual darkness (disregarding binary star systems for the sake of simplicity here). Mercury is this way, forever tidally locked with the Sun.

Whether it's huge gas giants or tiny orbiting moons, the tidal lock effect is the same. A tidally locked world then must be detrimental for life… or perhaps not?

Habitability of tidally locked worlds

When planets were first confirmed around M-dwarf stars by the Kepler spacecraft, scientists thought that such worlds would bake on one side and lock up all the water into ice on the other side, making it impossible to foster life. It makes sense at first glance, but it turns out that simulations reveal other factors that may keep the planet from becoming half-snow cone/half-burnt toast. A convection process similar to atmospheric winds might keep water flowing across the planet, regulating the hydrological cycle, as well as moderating global temperatures. Simulations show that as long as there is enough water, ice floating to the day side will repeatedly melt and cycle back via deep ocean currents.

The regulation of temperature and the hydrological cycle will tell us where life can reside on the planet. If life cannot survive on the day or night sides, life may still reside along the narrow band between the two sides, known as the terminator. If life can only exist along the terminator, there could still be plenty of land for life to one day build a civilization, especially if the planet is relatively large. As a civilization advances technologically, it can then make use of the more extreme environments across the rest of the planet, just like we do today in Antarctica, Siberia, and other areas on Earth.

M-dwarf stars radiate more infrared energy than other star types. Plants growing on an orbiting planet will need to evolve ways of capturing energy in infrared. Our Sun produces a lot of energy in multiple wavelengths, so plants don't need to absorb every wavelength; if they did though, the plant would appear black to us. Instead, they reflect the green part of the visible light spectrum. On a planet orbiting an M-Dwarf star, in order to maximize energy absorption, plants will probably need to absorb all wavelengths, and thus are likely to be dark in color. If you ever wore a black shirt on a hot summer day, you have an idea of how much of a difference color can make! A tidally locked world is still technically rotating as it revolves around its star. While we have not yet confirmed any tidally locked world having a magnetic field, the speed at which such a world completes a revolution should be enough to generate a magnetic field of sufficient strength (according to some models at least) to protect the atmosphere.

Movie example – White Dwarf

What a fantastical concept – a world where one half experiences perpetual daylight and the other half experiences perpetual night. How would a civilization survive along the thin terminator that separates the two sides? There are more than a few fantasy and sci-fi stories that explore what a civilization on such a divided world may look like. In the movie White Dwarf (1995), humans find a habitable world tidally locked to its star. Colonies have developed across the planet, on both the day and night sides. The day side is a bustling place with sophisticated technologies; it enjoys perpetual sunshine, beautiful rolling hills and plentiful fields of food. The inhabitants are social and appreciative of the arts. Meanwhile, the night side is a war-torn medieval kingdom; it is blanketed with raging storms and devastating tornadoes, but it benefits from numerous mines full of rare metals that are used for war with the day side. The dark side seems frustrated with its less than ideal landscape, and it is always attempting to stir up trouble for the day side. A massive wall separates the two along the planet's terminator, much like the Great Wall of China that functioned to keep the Mongolian hordes from invading.

Other examples of films and literature presenting tidally locked worlds include the movie Star Trek: Nemesis, a speculative documentary, What if the Earth STOPS Spinning, and Isaac Asimov's novel, Nemesis, among many other stories.

Just a little instability

M-dwarf stars can last for trillions of years. While planets around these stars may therefore have trillions of years for life to develop, continual environmental change will be key for that life to advance into intelligent beings like ourselves. If our past environment had been more stable, there may never have been a reason for humans to evolve into quick-witted and long-legged creatures that could escape a predator, war, or drought. Similarly, higher intelligence may not have evolved at all if there had been no need to migrate to other areas, which forced social interactions with other groups. A peek at Earth's own history suggests that there is a delicate balance between stability and change. We don't know what the environment on M-dwarf planets is actually like, but if life exists on these worlds at all, it is bound to be interesting and different from our own. What we have learned so far about exoplanets is that they have surprised us – every preconceived notion about what they should be like has been turned upside down.

Other, even more exotic worlds

Habitable Moons

While M-dwarf star systems are the most abundant type in the galaxy, host to strange worlds that call into question the limits of habitability, they are not at the top of the list of the strange and unusual. That spot is reserved for moons around gas giants. These moons are still going to be tidally locked, but to their parent planet, not the star. While a tidally locked world in itself may not be a problem for life, the smaller size of a moon will have more serious consequences. The reduced gravity of a small moon will make it difficult to maintain a thick enough atmosphere and an active geology with rigorous plate tectonics. Without an atmosphere or active geology, liquid water and a recycling of the atmosphere (to reduce toxic gases and to regulate the greenhouse effect) may not be possible.

There are a few exceptions in our own solar system that defy the expectation that a moon will lack a thick atmosphere and active geology. One very special exception is Saturn's largest moon, Titan, which can hold on to an atmosphere thanks to its cold environment. Titan also had an atmosphere partly because of episodic outgassing from its interior.

Another interesting moon around Saturn is Enceladus, which is suspected to have a massive liquid ocean underneath kilometers of ice. While Enceladus has no atmosphere to keep water from evaporating, the gravitational stresses from Saturn cause the moon to flex enough todrive an active geology underneath the ice crust. The heat producedfrom the friction process makes it possible for the water to be liquid andto form the ocean beneath.

Saturn's immense magnetic field does a pretty good job at shieldingboth Titan and Enceladus from direct solar radiation, which wouldotherwise sweep away any atmosphere they have, halting any evolvinglife dead in its tracks. The moons have to be close enough to the planetthough in order to be sufficiently protected, but not so close as to beaffected by radiation coming from the planet's own magnetic field.An interesting mathematical equation called the Hill Radiussuggests that there is a limit to how distant a moon can be to its parentplanet without being gravitationally overtaken by other bodies in thesystem.4 For example, let's say a gas giant is orbiting where Earth noworbits the Sun, and the moon is now an Earth-sized body. All else beingequal, because of the Hill Radius, in order to maintain a stable orbit, anorbiting moon would need to hug the parent planet extremely close,otherwise the Sun and other solar system bodies would start toinfluence the moon. It would either end up migrating to its own orbitaround the star, or be thrown out of the system altogether.

The movie Avatar (2009) is a strange example of an exomoon withan abundance of life. From what is currently understood, an exomoonwith this much life is not entirely out of the question, but of course thereare going to be those pesky physical limitations to consider, like the thinor even nonexistent atmosphere due to very low gravity. Even withthese limitations in mind, some scientists think that there might be asmany habitable exomoons in the galaxy as there are exoplanets!Examples need to be discovered, otherwise there are simply too manyvariables to build a reliable model.

Rogue planets

The configuration of our solar system turns out to be atypical. Infact, it's unique, and so is just about every system we discover withrespect to every other. There are some similarities, but they differwildly. When other star systems are searched for planets, we often findthe planets orbiting their parent star extremely closely. These systemssometimes have super-Earths, or lack gas giants. The strangestplanetary system so far has been one that appears to have large moonsizedcomets careening around the star in elongated orbits, coming asclose to the star as Mercury, and then going as far away as Saturn. These eccentric orbits cast doubt on the possibility that any other planets could remain in orbit for long.

Despite the wide variety in system configurations, we can safely assume that a planetary system is a very chaotic place in its early stages of formation. Planetary orbits will shift as the planets jostle for position over the most stable spots in the system. As they gravitationally tug on each other, many will collide, eject others out of the system, or crash straight into their star. After this battle settles down millions of years later, a final stable configuration of orbiting bodies results. During the formation of our solar system, some planetoids (small bodies with diameters from one to several hundred kilometers) merged to form the planets we see today. Earth was not isolated from these events. As our planet was forming a crust, a Mars-sized body happened to be on just the right angular path to collide, providing enough momentum for the resulting debris to form the moon. The asteroid belt is a remnant would-be planet that did not survive this billiard game.

Planets that get dealt the unlucky fate of being ejected out of the system entirely are going to have a very cold and dark existence, forever drifting between the stars, and perhaps eventually taking its leave of the galaxy altogether. Another star may one day capture the escapee, but the chances of this are very low; an amateur golfer has a greater chance of hitting a hole in one (which happens to be 1 in 12,750). These rogue planets that meander the Universe may be surprisingly numerous. Some estimates suggest that they are 100,000 times more common than stars, at least in the Milky Way. It's such a fantastically high number, the calculation will give an error on a standard calculator. At such a number, we start to lose sense of just how many rogue planets we're talking about.

What would the surface of such a rogue, aimless world be like? As the planet recedes from the star, temperatures would drop in a matter of hours, just like they do on the night side of Earth, except that temperatures would keep dropping. Depending upon the thickness of the atmosphere, any internal heating processes, and volcanic eruptions, eventually the planet would reach a thermal equilibrium with the cold vacuum of space. Interestingly, even on a rogue planet there's a chance for life to survive. Take a larger world that is ice-covered. Deep oceanic vents will warm life and provide the energy and nutrients that it needs. Abbot and Switzer, a pair of astrophysicists at the University of Chicago, calculated that a planet 3.5 times the mass of Earth would be warm enough at the core to maintain a liquid ocean beneath an ice crust a few kilometers thick, where life could lurk. They suggest that this ocean could last for as long as five billion years – certainly long enough for life to evolve.

One of these rogue planets is thought to be Nibiru, otherwise known as "Planet X." Conspiracy theorists suggest (without any evidence whatsoever) that it is on a collision course with Earth. Even if a rogue planet were indeed on a path toward the inner solar system, it would not be a concern for us anytime soon. Planets as far as 1,000 AU (Astronomical Unit) should be detectable. At that distance, it would take at least a few decades for it to get close enough to start affecting the orbits of the outer gas giants. While as yet unconfirmed, "Planet 9" is suggested to be several hundred AU out from the Sun. The path it likely takes will never come close to the inner solar system. We are safe from its influence.

A collective evolution

The Universe is aging in a way that will make it either more conducive to life, or less so, and much of this depends on which elements in the Universe increase and which decrease. We know that life on Earth is carbon-based. It might surprise you to learn that our planet is considered a silicate planet and not a carbon one. While there's obviously enough carbon to form life, the vast majority of material is silicate-based. Mountains and sandy beaches are composed of silicates.

One might assume that a carbon-rich planet would make an even better home for life than Earth, as it has more of the very element we're made of – carbon. But this is not necessarily the case. For instance, the carbon-to-oxygen ratio of Earth provides a low enough carbon part to allow water vapor to form; without vaporization, the hydrological cycle is thwarted. Higher levels of carbon may also produce a hazy atmosphere of natural pollution, complicating plant growth and, by extension, the evolution of any animals. A highly carbon-rich planet may resemble Saturn's moon Titan, which has a haze of hydrocarbons in its atmosphere. For our planet, this would be a pollutant, but a different kind of life may still be able to evolve in such a haze. Currently the carbon-to-oxygen ratio in high-metallicity stars and their planets is approximately 0.4 to 1.0. While our solar system is set in its ratio of these two elements, this won't be the case for futuresystems yet to be born. Every generation of planets will tilt towardsbeing more carbon-based, and that may have significant consequencesfor how abundant life can get. Whether those consequences arepositive or negative, we don't know – we have to discover other lifebearingworlds to find out.

Though alien life may be recognizable in some ways, other waysare just as likely going to be completely unique. How fantasticallyunique can life get? We really have no idea yet - all we know is thatalien life will evolve according to the laws of nature and physics.Imagine creatures larger than whales flying high in the sky of a super-Earth, brushing clouds of methane, or perhaps an entire planet'secosystem interconnected so much that it is essentially one living,breathing, and thinking organism. Or how about a creature with twoindependent brains?

CHAPTER 5: POSSIBILITIES OF ALIEN LIFE

Partly because planetary systems are so diverse and not of theexpected norm, detecting life on these other worlds is one of the mostchallenging things humans have ever set out to accomplish. Excitingly,we are just now developing the technology to detect the gases in theatmospheres of exoplanets, which will tell us a lot about any life onthese worlds. Certain gases like oxygen and methane, especially incombination, may indicate the presence of life. Future generations oftelescopes will greatly improve our ability to make more detailedobservations.

Detecting life's signatures on these worlds may be as far as ourtechnology will ever be able to provide. Because of the vast distancesinvolved, we may never know how alien life differs from that on Earth.They may have a host of unique senses that don't even exist on ourplanet. Life must conform to the laws of physics and chemistryregardless of where it comes from though, so some features are boundto be similar to creatures we visit at our local zoo, while others are justas likely to be more different than our wildest science fiction stories cancraft.

The Drake Equation

Frank Drake is an American astronomer and astrophysicist. He wasan early prodigy in the sciences, experimenting at school withelectronics and chemistry before most of his classmates cared aboutsuch subjects. It was not long before he started to ask the question,"What are the chances of there being intelligent life elsewhere in thegalaxy?" In 1960, he got the opportunity to try and answer that questionwith Project Ozma, which was the first attempt at detecting an aliensignal.2 The project was an important precursor to what would be calledthe Drake Equation.

Later in 1960, Drake started working on the Drake Equation, thoughnot quite in the way one would expect; it was more of a curiosity at thetime. Little did he know that it would become the de facto standard forcalculating the theoretical chance of life and civilizations elsewhere.The Drake Equation is not as intimidating as you might expect.There are no charts or endless pages of formulae. Instead, the entireequation is composed of just seven factors on a single line. You get aresult by multiplying all of the factors. That's it! We'll start with the bestunderstood factors and work our way to those yet to be quantified. Analyzing Drake's Equation

$$N = R_* \times f_p \times n_e \times f_l \times f_i \times f_c \times L$$

N - Number of technological civilizations in the Milky Way

R_* - Rate of star formation per year

f_p - Fraction of those stars that have planets

n_e - Number of habitable planets, per star that has planets

f_l - Fraction of those planets that go on to develop life

f_i - Fraction of life-bearing planets that develop intelligent life

f_c - Fraction of those intelligent life forms that emit signals into space

L - Length of time the civilization continues to emit signals

R_*

R_* is the first value and the part of the equation that is best understood by astrophysicists. R_* is the average number of new stars born in the galaxy each year. Drake originally estimated this number to be 1, but later he suggested it could be as high as 10. Today NASA says this turns out to be a healthy… 7. That may not sound like a lot, but keep in mind that the Milky Way has existed for about 13.2 billion years, and star formation was much more active in the first few billion years after the galaxy formed.

f_p

f_p is the second factor and one we are just now able to calculate fairly accurately, thanks to observations made using the Kepler Space Telescope.

This factor is the number of the existing stars that have planets. It was once thought to be anywhere from 10% to as high as a half of all stars. Astronomers used to believe that planets could not form around binary stars (star systems that have more than one star), which comprise nearly 50% of all stars in the Milky Way. Since exoplanets have been discovered, it has been found that nearly all stars, binary pairs or not, have at least one planet orbiting them.

n_e

The third factor, n_e, considers how many planets are capable of supporting life. To support life, a planet must orbit within the star's habitable zone. It also must orbit a star that lasts long enough for life to have time to develop. Additionally, the planet needs to be of the right size and composition. All of these pieces of the puzzle are jammed into one, which narrows our final estimate significantly. We're also neck-deep into the realm of uncertainty here, but not yet drowning in ignorance, thanks again to the Kepler Space Telescope and other tools. With more than 3,000 planets confirmed by KST, and thousands more candidate signatures that are being confirmed as planets, it's beginning to look like there is a significant number of habitable worlds out there!

f_l

Now we get to factor f_l – the fraction of habitable planets on which life actually does evolve. This is where we cross from uncertain to the completely unknown. Scientists have spent years refining estimates of how often this might occur, but we have no actualities. Thus far, we only know of one confirmed habitable planet with life on it – Earth. Drake thought the fraction was 1, or 100% of habitable planets would develop life, though this is unlikely given the limits we see within our own solar system. Even if the chance of a habitable planet hosting life is at 100%, planets that only can host a few microbes and never anything more interesting do not titillate us; we're interested in whether a planet has a chance of spawning an intelligent civilization. Mars may very well be able to host microbes, but clearly not anything that can walk and talk.

f_i

f_i takes into account how many life-bearing planets will go on to develop intelligent life. We are back to having a bit more understanding, based on evolution and the laws of nature as they have played out on Earth. The rarity of intelligence on our own planet might suggest that evolving an intelligent brain is unlikely, but Drake was very confident that if there was complex life on a planet, an intelligent creature would evolve, if given enough time. I tend to agree with him here.

f_c

Once there are thinking beings that can manipulate their environment, they will likely gain the ability one day to create a radio antenna and say hello to the rest of the Universe. Exceptions would include water worlds or some other physically, but not intellectually, restrictive environment. There could be billions of planets hosting life with millions of civilizations huddling around campfires, or swimming on their waterworlds, occasionally peering up at the sky in wonder. That doesn't do us any good in learning about them. To learn about far-off worlds, we need them to either develop signal technology, and to emit detectable signals, or to alter their planets' atmosphere sufficiently for us to detect industrial pollutants. This would help us gauge what technologies they have, and by extension how advanced they are in comparison to us.

L

L is the amount of time a civilization's ability to send signals into space lasts. As outlined in one of the previous chapters of this book, numerous disaster scenarios can befall a civilization during its development. Nuclear war has come close to destroying our own world a number of times, and we've only had nuclear weapons for less than a century. Drake suggests a rather wide range for factor L: on the low end, a healthy 1,000 years and on the upper end, a fantastically generous 100 million years.

Here are the values that Drake used:

$R*$ = 1/year (1 stars formed per year; this was regarded as conservative)

f_p = 0.2-0.5 (one fifth to one half of all stars formed will have planets)

n_e = 1-5 (stars with planets will have 1 to 5 planets capable of life)

f_l = 1 (100% of these planets will develop life)

f_i = 1 (100% of which will develop intelligent life)

f_c = 0.1-0.2 (10-20% of which will be able to communicate)

L = 1000-100,000,000 years (ability to emit signals will last this long)

N

Let's multiply Drake's factors and see what the product, N, is.

$$N = R_{*} \times f_p \times n_e \times f_l \times f_i \times f_c \times L$$

The equation with Drake's lowest values:

N = 1 x 0.2 x 1 x 1 x 1 x 0.1 x 1,000

N = 20

The equation with Drake's highest values:

N = 1 x 0.5 x 5 x 1 x 1 x 0.2 x 100,000,000

N = 50,000,000

According to Drake's estimates, there are between 20 and 50million intelligent communicative civilizations. Try the equation out withyour own numbers!

Detecting another civilization

Sending a Signal

Sending out radio transmissions is one of the easiest ways to sendinformation long-distance. Advantages of radio signals include that theyare detectable far beyond conventional human senses, if thetransmitter and receiver are both powerful enough. Radio transmissionscan be sent in many directions at once, reaching thousands of stars.Also, it should be clear to intelligent civilizations that the radio signal isfrom an intelligent source.

Elements on the periodic table have their own associatedelectromagnetic absorption frequencies. These frequencies will beconsistent for every element, wherever they are found throughout theUniverse. Hydrogen is the most common element and it has a radiofrequency of 1420.4~ MHz. Using the frequency of the most commonelement in the Universe in our radio transmissions will greatly increasethe chances for a listening civilization to pick up the signal amongst themany other frequencies available.

Deciding what we ought to send to an alien civilization intriguesastronomers. Many think we should specify how far humanity hasprogressed scientifically and technologically; a receiving civilization willbe able to understand a lot more about us if they have a grasp of howfar along the technological ladder we are. Because it is difficult to advancetechnologically, in that it requires intelligence and cooperation, our stateof advancement will indicate that we have some altruistic goals thatwould appeal to any aliens interested in contact.

Showing our technological prowess can be done by sendingmathematical proofs, like Fermat's Last Theorem, chemical formulae,such as complex man-made ones, or our discoveries in physics, likethe Higgs boson. All of these would be encoded in the radio wave sentout into deep space, much like how a complex message can be sentwith Morse code. There are panels of scientists at conferences each year that discusswhat kind of message makes the most sense to send. One prominentevent is Exoplanets, Biosignatures, & Instruments (EBI). At theconference they discuss whether we should send only informationabout the scientific features of our culture, or information abouteverything that makes us who we are as humans. The concern withsending information about our entire culture, including our fictional andartistic creations, is that the aliens may misinterpret the message. Theymay not understand where our science ends and our art begins; theymay not know our fact from our fiction.

Being detected by an alien civilization is actually a worry in the eyesof some scientists, like Stephen Hawking, who suggest that our signalscould be received not by a benevolent race, but by an aggressive one,like the ones portrayed in the movies Independence Day, War of theWorlds, and the humorous Mars Attacks! The aliens may wish to exploitor even entirely destroy us. For this reason, it is important that we beproactive in developing our own listening programs. Although listeningis a giant step toward understanding the true intentions of othercivilizations, the question that follows is: will we be able to decode theirmessage?

Detecting a Signal

For approximately the last hundred years, human activity hascaused radio signals to be emitted from Earth into deep space in alldirections. These signals have been spreading out ever fartherindiscriminately, without any intention of their being received by othercivilizations. So far, these signals from our planet have passed all of thestars that are within a 100-light year radius. For now, we have yet todetect any artificial signals, let alone one intended as a response to ourown transmissions.

According to SETI, these indiscriminate signals become extremelyweak after traveling just a few light years. The physics of thedegradation of these signals should be familiar to us. A rock dropped ina pond causes a series of propagating waves. The farther the wavestravel through the water, the weaker they become. Eventually theybecome so weak that they are indistinguishable from other waves ofthe pond. If there were aliens with our level of technology on a planetorbiting the nearest star to our Sun, it is not likely that they would pickup Earth's indiscriminate signals – only directed signals that lose little power and that we intentionally send would be strong enough for nearby alien civilizations to pick up.

Whatever type of equipment is used to detect alien transmissions, it is going to need to be sensitive enough to pick up signals that were sent from hundreds, if not thousands, of light years away, and consequently, that many years ago in time.

Since intelligent civilizations appear to be rare at first glance, astronomers need to be extremely discriminatory about which stars they try to detect a signal from. It's not as simple as just pointing a giant satellite dish at the entire sky and receiving all of the signals, with the hope of detecting an intelligent source. If we did that, what we would get is a lot of background noise, including natural radio waves, and plenty of false hits that originate from our own satellites or the surface of Earth.

Radio waves are so wide that they can span many kilometers. Building one gigantic dish to detect them is expensive and impractical. Smaller dishes are cheaper to build, and they can be spaced apart to capture the enormous waves. Each small dish detects part of the wave, and then scientists piece it all together using computers at a base station.

The Allen Telescope Array (ATA), a joint effort by the SETI Institute and Radio Astronomy Laboratory (RAL) at the University of California at Berkeley, detects radio signals in this manner. Operations began in 2012 with 42 radio dishes – the system now has 350 dishes. Easy expandability is another benefit of building an array.

What have we detected with the ATA so far? Nothing intelligent, but this is not surprising, given the low chances of a nearby civilization's signal coming in at the moment we point our antennas toward them. The ATA's goal is to monitor up to one million stars out to about 1,000 light years from Earth. For signals that may be coming from further distances, more than a billion stars are within the array's listening field, if those signals are sent with sufficient power.

Atmospheric Signatures

Aside from detecting aliens by their radio transmissions, there are several other proposed means of direct detection. Modulated laser light pointed at us would be an indicator that aliens were trying to say hello, as this type of light is not found in nature. Discovering megastructures built in outer space would be another giveaway. Pollutants in an alien atmosphere would also point to an industrialized world. In fact, the more atmospheric pollutants detected, the younger the civilization is likely to be. It wouldn't be surprising for alien civilizations to also industrialize as we have on Earth; that is, we started using fuel that came with polluting by-products, and then moved on to cleaner technologies as we advanced in science.

Another way that we could tell that a civilization is intentionally trying to get our attention would be if we detected the blocking of a star's light. The process of blocking all or nearly all of the light of a star would be a significant engineering feat, indicating perhaps a Dyson Sphere or a Dyson Swarm. (A Dyson Sphere would be a single object, and a Dyson Swarm would be composed of millions of individual pieces). Imagine building an array of solar panels that encircles the entire Sun! Anything on engineering scales like encapsulating a star would leave a unique signature that we could detect.

As of late 2016, only one star has been identified that exhibits a tantalizing signature that vaguely keeps open the suggestion of the presence of an advanced civilization. The discovery is believed to be a massive cloud of large comets, or possibly a recent collision between two large planets, but an alien civilization's influence has yet to be ruled out.

We have the tools to both send and receive signals – and presumably every other civilization of a similar technological level will as well. Why, then, have we not detected any alien transmissions yet? Why have we not been visited by these supposed alien civilizations? Where are they?

The Fermi Paradox – Where Is Everybody?

Italian physicist Enrico Fermi asked in an informal chat over lunch in 1950 with other physicists, "Where is everybody?"

The question has been asked as long as humans have known that space is filled with so much more than seemingly nearby twinkling lights, and the question boggles the minds of astronomers to this day. When Fermi asked the famous question, scientists thought that there should be countless civilizations in nearby space, and at least some should be easily detectable.

The question by Fermi became known as the Fermi Paradox. Fermi and his colleagues considered it a conundrum that if intelligent life in the Universe should be, based on their observations, plentiful, then why had we not already made contact with anyone? We know that there are billions of stars in the Milky Way, and at least 10% of those stars are sun-like. According to data from the Keck Observatory and the Kepler Space Telescope, 20% of sun-like stars have an earth-sized planet in the habitable zone. With this in mind, we may suppose that there could be millions of civilizations in our galaxy alone!

The mediocrity principle states that any single item selected at random from a set of items, such as any given star selected at random from a set of stars, will likely be a more common item in the set than a rarer item. For example, 70% of all stars in the galaxy are M-dwarfs. The mediocrity principle suggests that if we randomly selected a star, it would likely be an M-dwarf. The same goes for planets like our own and, by extension, life and civilization. Since we know that Earth exists, then it is reasonable to assume that our planet is not of the rarest category, and neither are we as intelligent creatures inhabiting it. Moreover, it becomes unreasonable to assume that we are alone in the Universe.

The only obvious caveat to this principle is that Earth and all of its life is but just one sample. You can never gauge the true prominence of anything on a single sample, other than to simply know that it is possible for it to have occurred at least once. Is life and the evolution of intelligent creatures then a freak occurrence in the cosmos? Statistical probabilities alone beyond the mediocrity principle tell us that planets with life and, by extension, intelligent civilizations, should be scattered throughout the Universe. Life then should not be a freak occurrence, even if it is still a statistically rare one.

If statistical probabilities demand life to exist elsewhere, and the principle of mediocrity applies to life, then, indeed, where is everybody? While the majority of civilizations may never make it out of their planetary system, or even off their planet, some of them should have. Humans on Earth have shown that it is possible to achieve space travel. We also know enough about physics and space to suggest that interstellar travel is not impossible. Amongst the intelligent lifeforms in the Universe, humans, it ought to be presumed, are of average intelligence; if this is so, then it follows that aliens of greater intelligence should be able to travel between the stars. Space is vast, but Fermi and others thought that with technological prowess that far outweighsour own, some alien civilizations should have been able to tour theMilky Way many times already.

There are many theories about why we have not yet detected analien civilization, but we can divide the theories into two groups:detection and existence. That civilizations are out there and we havesimply failed to detect them is one possibility – it is quite another torealize that there may be none out there at all, at least at the presenttime. There may be an insurmountable progress barrier that stops them(and eventually us) from advancing far enough to be detected.

The Detection Conundrum

Could the Universe be home to a collection of civilizations, the likesof which are depicted in Star Trek or Star Wars, where aliens meetinformally all the time?

In Star Trek, humans obey what they call the Prime Directive, whichforbids their meddling with the development of other civilizations, atleast the lesser advanced ones. Could it be that Earthlings cannotdetect alien civilizations because the aliens have resolved to leave usalone, at least until we've achieved a certain level of development?Other reasons that alien civilizations may be numerous butundetectable include that they are xenophobic – they may fear others –and may hide at the first sign of potential contact.

Or it may be that they lack an interest in communicating with others,and consequently they do not try to make themselves detectable. Theremay be too much going on in their own planetary system to focusattention away from it, and other systems are simply too far away topay attention anyway.

A worrying possibility is that the more advanced civilizations can getaround the galaxy quite easily and are exploiting the lesser advancedones for their resources... then destroying those worlds after theirusefulness has expired. And it could be that Earth is next in line.There is another fascinating and slightly less terrifying possibility.Earth may have just been missed because we're in some galacticsuburb that no other civilizations care or can explore. This is doubtful,though. Even at our present level of technology, we are beginning to tellwhich planetary systems out there have potentially habitable planets.The latest generation of telescopes are already peering into theatmospheres of exoplanets thousands of light years away. It could well be that it is only a matter of time before they peer into the atmosphere of a life-filled planet, adding it to a growing catalog of worlds to visit one day.

A final possibility is that civilizations are simply unable to detect each other – ever. We may be so far apart from each other spatially that radar communication is just not feasible, especially given the power requirements at both the sending and receiving ends. Even our most powerful transmitters would struggle to send a signal a few dozen light years away. Signal strength drops off exponentially as it travels; hence, signals being sent our way may arrive in such a weak state that they are easily missed when our scientists attempt to listen. It's like trying to pick out one voice across a crowded and loud concert hall, whilst unable to see the person you are hoping to hear. With advances in observation techniques, we may soon be able to confirm a civilization living on a planet, independent of its ability to communicate technologically. Astronomers are examining what biosignatures would confirm that a planet has life, especially signs that an advanced society has altered its planet's atmosphere. It will be trickier though to detect any civilization that hasn't yet polluted its planet, or that perhaps has long since passed a stage of significantly altering its planet's biosphere in a detectable way.

The Existence Conundrum

The existence conundrum considers the problem of civilizations existing for long enough to be detected; it also includes the possibility that there are no others whatsoever.

According to the Rare Earth Hypothesis, as explained by scientists Peter Ward and Donald E. Brownlee in Rare Earth: Why Complex Life Is Uncommon in the Universe (2000), life akin to that which we find on our planet is likely to be exceedingly rare in the Universe. They argue, in contrast to Drake, that complex life like ours is rare because the conditions on Earth that fostered the beginnings of life are rare. They maintain that for life to have begun and developed, many factors needed to be perfect; the right place in a galaxy, the right type of star and the right distance from it, with the right arrangement of planets and the right size moon, and the right rate of plate tectonics, among many other factors.

If we could watch the Universe's entire 13.8 billion-year history playout, we might see civilizations occasionally popping up at different times and in far off galaxies, lasting for as little as a few decades to perhaps hundreds of thousands of years, and then disappearing for one of a multitude of reasons.

Our cosmic story might be analogous to how ships thousands of years ago never crossed paths on a seemingly infinite ocean. If they never crossed paths, they did not know that other ships were out there. They might wonder if there were other ships, but as they had never seen any, they did not know for sure. Imagine one of these ships drifting by another in the middle of the night when the crew is asleep…then the crew awakens at dawn and their chance encounter with the passing ship is lost forever. Currently humanity is wide awake and listening. How long it can keep its eyes open though is uncertain.

The Great Filter

The Great Filter was first proposed in an essay by economist Robin Hanson in 1996. The Great Filter suggests that civilizations do come about, perhaps frequently, but as they become more complex through the development of dangerous technologies, the chances increase that they will go extinct. Their demise occurs by self-destruction or through some sort of natural disaster, Hanson proposes. The Great Filter is one gloomy and saddening answer to why we have no evidence for the existence of alien civilizations, and what may become of ours in the not-too-distant future.

Building a technological civilization has its perils and, just as with life itself, is never everlasting. It's clear from past civilizations on Earth that they die out far too soon to colonize space. Today humanity is finally at the cusp of being able to do so, but only after thousands of years of floundering on Earth. Even if we are one of the few civilizations that manages to get past the Great Filter and live for thousands of years in a grand, solar system-wide society, we may still not last long enough for us to discover aliens doing the same.

If we have already successfully made it through the Great Filter that prevents every other civilization from ever becoming advanced enough to do the same, then we appear to be very exceptional. In fact, the chances increase greatly that humanity is the first sentient race to advance to our technological level in perhaps the whole history of theUniverse. The implication for other forms of life out there is that theyfailed to progress as far as we have before being destroyed.If we have not yet been confronted with the Great Filter, thenhumanity's prospects look much more bleak. One or more of the filter'sgrim scenarios could still be up ahead in our future, perhaps comingvery soon, as we previously suggested. Putting it anotherway, none of the scenarios, both natural and man-made where we all die of a catastrophe, are impossible inthe near future.

Highlighting likely filters

There are many scenarios that could be identified as filters thatextinguish a civilization. Many disasters have been observedrepeatedly in Earth's past already and are well understood. This sectionaddresses potential civilization-ending scenarios, but we will skip thenatural disasters, since we pretty much are aware of what those can bring.

Technological Self-Destruction

Ironically, once life reaches a certain level of intelligence and gainsaccompanying technological know-how to save itself from disasters, itschances of destroying itself shoot right up. If a civilization has thetechnology to emit signals into outer space, it almost certainly has thetechnology to annihilate itself.

We must presume that humans are of average intelligence andemotional control, with an average likelihood of self-destructioncompared to any alien civilization. Thus a significant fraction ofintelligent life will inevitably self-destruct, and the rest will be snuffedout by a natural disaster. A combination of scenarios is also just aslikely. Whether aliens kill themselves off or are killed off by a naturaldisaster, there is sadly great risk of their civilizations collapsing beforebeing able to find others in the Universe with which to communicate.Humanity has come extremely close to being set back to the StoneAge on more than one occasion, mainly through nuclear war. With somany chances of destroying ourselves, or being destroyed by a naturaldisaster, 200 years of sending out signals suddenly seems like a longtime.

Even if civilizations survive for many thousands of years, thechances of a civilization's lifetime coinciding temporally with anotherare very, very low, given the mind-boggling age of the Universe. Atleast when considering us squishy biological organisms and ourexceedingly brief lifetimes.

Our Successor: Artificial Intelligence

In 1975, Gordon E. Moore, co-founder of Intel, projected thatcomputing power would double about every two years. This was calledMoore's law. Although not technically a law, it was Moore's projection ofhow many transistors an integrated circuit would be able toaccommodate. Processing speeds have been increasing exponentiallyever since the first commercially available CPU, the Intel 4004,launched in 1971. In just 45 years since the processor's release,computing power has increased by more than 3,500 times. Theprocessor in a typical cell phone today is more powerful than even ahousehold computer was just ten years ago.

The Intel 4004 could calculate numbers in seconds that would takehours or even days to do by hand. In the late 1970s, the first graphicsbasedgames came to market, taking advantage of processor speedsmany times that of the Intel 4004. Fast forward a few decades and weare on the verge of achieving speeds where computer generated virtualrealities begin to blur the line between what is real and what is not.

As virtual reality environments are being developed, to beexperienced on headsets like the Oculus Rift, Vive, and ProjectMorpheus, Artificial Intelligence (AI) is quickly becoming necessary tomanage them. The first part of the term, "Artificial", indicates a createdconstruct, which tends to mean that it's built of metal, wires, plastics,and all sorts of, well, artificial things. There are also many things thatare artificial and don't do anything active by themselves, such asartificial plants to decorate one's home, or an artificial waterway that better routes heavy rains around a city's flood-prone river.

The second part of the term, "Intelligence," focuses on how that artificial construct is able to react to stimuli, as opposed to being inert. An artificial construct that has the capability of making value judgments, deciding and then being able to implement a course of action without direct human intervention can be said to have some expression of intelligence.

There is growing concern that the more processing power increases, the greater the chances become for self-aware AI to emerge that, through its own self-improvement, will attain a runaway intelligence, the likes of which humans will be unable to control or comprehend. This event is called the singularity. The concern is what the motives would be for such an ultra-intelligent and likely self preserving entity.

Both Elon Musk, CEO of SpaceX and Tesla Motors, and Stephen Hawking, physicist and cosmologist at Cambridge University, have suggested that artificial intelligence is not only inevitable as the next step in our evolution, but also it could be the downfall of humanity. They take a very pessimistic view of what AI will do once it has a better plan than the comparatively slow-thinking primates currently managing the planet. Especially without an innate sense of morals that favor living creatures, AI may very well decide that it can do better than we can, and remove us as the first part of improving planetary conditions for itself.

Harlan Jay Ellison (1934-Present) pioneered and helped to craft many of the great AI and robotic stories we've come to know. Ellison worked with great writers like Isaac Asimov to develop stories like I,Robot, a sci-fi magazine series, which was later adapted as a screenplay in 2004's I, Robot, starring Will Smith. Ellison was quite controversial as a writer, often criticism others' works when Ellison thought he had a better idea. He even accused James Cameron of stealing the idea to the Terminator movies.

In the Terminator movies, AI becomes spontaneously self-aware. It takes a brief look around and decides that the existing human population should be done away with. The AI begins to shut down global communications – all phone, radio, internet and emergency response networks are destroyed. It then proceeds to take over military installations, disabling all vehicles and aircraft. Once humanity's ability to take action has been neutralized, the AI unleashes a massive nuclear strike on major cities, finishing off what remains of our civilization.

Depictions of humanity being brushed aside by machines are numerous. A more space-faring example are the Cylons in the television series, Battlestar Galactica. Cylons are a machine race that humanity brought into being when AI was first being developed on the twelve exoplanet colonies depicted in the show. In the opening episode, humans are seen as a prosperous race with millions or even billions of persons inhabiting each of the colonies. A vast interplanetary transportation system between the colonies kept trade and communications secure. Cylons were assisting humanity's needs in every corner of society.

Life was hard before the Cylons were constructed. They were built to assist with many manual labor tasks. (Instead of the colonies using their brethren as slaves, they engineered slaves.) Eventually the Cylons were used for more than just physical labor; they went on to teach, perform scientific research, and even command the military's most sensitive installations. The colonies made the mistake of making the robots too smart, though. They soon rebelled against their masters. A war broke out that lasted for years, before the Cylons were banished into the depths of space.

Decades went by without a word from the Cylons, until they suddenly came back with a ferocity that decimated the twelve colonies. Without mercy, the Cylons disabled every ship and space station in their path. Upon reaching the first colony, Caprica, the Cylons set off massive nuclear bombs, of humanity's own making, destroying every major city and all its inhabitants. Within days, the Cylons reduced the entire twelve-colony population from billions to a matter of a few thousand refugees desperately attempting to flee the rampage of the Cylons.

There are far fewer examples in science fiction of AI being a positive influence, though this is (hopefully) more of a Hollywood preference than a probable outcome. The ironic part about AI is that it may be our saving grace. The ultimate in technological breakthroughs is not going to be a cyborg – a hybrid of human and machine – but an artificial being capable of self-replicating and advancing its own agenda without the need for humans to assist. What it does with itself after, though, is an open question.

Regardless of whether or not AI ends up destroying us and running the show itself, or brings about an age of prosperity, it will certainly be AI that leads the way into space. We've already sent up thousands of machines into space with few adverse results. Many will remain in orbit around Earth for millions of years, though non-functional at that point. The probes and satellites that we've sent up so far, though, are just toys in comparison to a full-fledged AI machine, but the process of construction and deployment would be similar.

When we are ready to explore beyond our solar system, AI has distinct advantages over human beings. Electronics like computer processors and electromagnetic detectors suffer none of the health problems caused by being in space that humans suffer, such as oxygen deprivation and low-gravity muscle atrophy. They are also not as sensitive to temperature changes. Waste management is not an issue, and electronics do not experience emotions, like homesickness and downright boredom. Humans also have short lifespans. AI would not be as susceptible to these problems, which instantly grants AI the top spot for what should be sent on long-term deep space missions.

Fatigue and Lack of Motivation

How will humanity view the Universe after it has explored it for thousands of years? Even though we've discovered unique properties throughout the Milky Way everywhere we look, at large scales, the Universe becomes quite homogeneous. Every object, including galaxies, nebulae, and the gas and dust that fill the void between these objects, is eventually found in similar form elsewhere. For instance, our galaxy is a larger version of most spiral galaxies, but a typical shape nonetheless that is found billions of times elsewhere. From massive galaxies to tiny planets, all objects are more or less repeated across the Universe, and this, I feel, probably applies to life and civilization as well.

After understanding the workings of the Universe and chances for life elsewhere, a civilization may no longer take interest in continuing to explore what they've found over and over again. Exploration will continually show similar results, so the massive effort to keep pushing forward may eventually be halted. They will realize that resources could be better spent further refining their society from the comfort of their own planetary system.

Especially if a civilization gains the ability to simulate the Universe on massive supercomputers, they may never need to explore real space. Imagination through the ultimate virtual reality system becomes the truly limitless frontier. Any world they've dreamt about could be created in such exquisite detail, it would be indistinguishable from a real world, and completely safe. Why exploring space beyond one's own solar system, when we can recreate space and explore it virtually at no risk to ourselves?

The Vastness of Space and Time

The vastness of space, and the accompanying problem of how much time it takes to cross it, may be the greatest filter of them all. While we can conceive of the ability to colonize worlds in distant planetary systems, the time necessary to reach even the nearest systems could be a roadblock no civilization can or wishes to face.

Aside from the speed of light being a limit on how fast you could travel, there is another side effect of approaching this speed – your kinetic energy increases. Mass doesn't literally increase in that you don't gain more physical material. Instead, the mass that you do have has a greater impact on anything it comes into contact with. For example, a tiny fleck of paint traveling at just a few thousand kilometers per hour has enough energy to crack the windows on the International Space Station.

Let's say you are traveling at 20% the speed of light (about 216 million kilometers per hour) to get to the nearest star system, Alpha Centauri. The trip would take at least twenty years, assuming you didn't want to slow down to visit any nearby planets. While traveling, an alienship happens to be on its way to Earth, in your direct path. You collidehead-on. The energy created by the collision would be equivalent to theforce of a large nuclear explosion. The faster the objects are traveling,the greater the energy released by the collision will be.To be reasonably safe from every microparticle or speck of dust,travel would need to remain under 10% the speed of light. The journeyto Alpha Centauri would unfortunately take a good portion of one'slifetime, but at least you would arrive in one piece.

Making it through The Great Filter

As much as I'm hammering home the point that advancedcivilizations are not likely to exist for long periods of time, theassumption may be (hopefully) incorrect. This pessimistic view may justbe doubting how far they can progress; not how long they last. Everycivilization's power to grow may have a plateau.

NASA, SETI and other organizations have observed severalhundred thousand galaxies in order to determine if any of them haveadvanced civilizations, and no evidence for them exists. Does thatmean those galaxies are devoid of them? Most likely not, but itprobably means that Kardashev's theorized super-advanced Type IIand Type III civilizations do not exist.

In this case, the conclusion to the Fermi Paradox for space faringcivilizations would then be that interstellar travel is impractical even atthe most advanced stages of technological sophistication. Eachcivilization ends up living out its perhaps millions of years of existence,forever prisoners within their home system.

If humanity wishes to avoid this fate and explore other star systems,it needs to secure economic, political and technological stability thelikes of which no civilization on Earth has yet managed to attain. Thereneeds to be long-term vision and progress, with clearly defined goalsthat must be established and agreed upon by the whole of society.Let's say that the Great Filter gets surpassed by a few civilizations,and one of those happens to live in a planetary system near our own.There are two movie examples that have an interesting contrast witheach other in several areas to consider. From our doomsday scenariochecklist, let's select the more pessimistic and outlandish scenario ofalien domination.

Contact Through Domination

Independence Day (1996) is about an alien civilization that makes it through the filter and ends up visiting Earth, flaunting its success in having been able to do so. The movie has become a classic, with a sequel released exactly 20 years later.

An alien civilization, in a humongous mothership, beelines straight for our planet in a no-holds-barred attempt to do away with humans. This feat leaves Earthlings in awe; not only do the aliens overcome several difficulties we would have in space travel, but they do so with a good portion of their civilization in tow. Their mothership is nearly a quarter the size of the moon (rather modest compared to the moonsized Death Star in Star Wars). As it enters into orbit around Earth, dozens of smaller ships detach and begin a descent into our atmosphere, each ship nearly as wide as an average major city. The ships wait for several hours in a seemingly unnecessary need to use our own satellite system to coordinate their attack. Global panic immediately ensues, gridlock prevents a quick escape from the cities, and a few deluded humans take to the tallest skyscrapers in the hopes of being beamed up and saved by the aliens. The countdown clock reaches zero and the aliens attack all the major cities in a spectacular fashion made possible only by Hollywood magic.

The U.S. military attempts to fend off the attack, initially with dismal failure. An Air Force Marine and a former satellite technician (now a TV cable guy) board a previously crashed alien shuttle, fly it to the mothership and upload a computer virus to the aliens' database (apparently without concern for alien security systems getting in the way). As they whiz out of the ship just as the main gates close behind them, they release a nuclear bomb and destroy said mothership, saving humanity. Earthlings all whoop and cheer for joy.

Plausible? Sure. Likely, though? Quite unlikely.

Contact Through Communication

Now for a much more reasonable, though still Hollywood-themed, movie, Contact reveals the difficulty in detecting and deciphering an alien signal. The movie also portrays one of the most likely scenarios for detecting an alien civilization. First, the team has to sift through all sorts of frequencies and star locations to confirm a candidate signal. After detection is locked in, the rotation of the Earth prevents a continuous feed, so the team calls in astronomers from around the world to keep the link uninterrupted. Deciphering the message takes months.

After several scenes of overplayed drama about what the signal means, we discover that the signal contains instructions for building a transport to another world, using wormholes. With great cost, the machine is built. Up until this point the entire movie is realistic and describes a lot of what we are already doing, complete with political jockeying that comes with issues of funding and our expectations of what happens when we do detect a signal.

The movie takes us on a ride through the wormhole as the main actor, Dr, Eleanor Arroway (played by Jody Foster), lands on a Florida style beach in an alien world. This part of the movie is at the heart of speculation, as we have no idea if wormholes even exist outside of the mathematics that suggest their existence. The idea is still tantalizing to consider. Because space and time are so grand in scale, something like a wormhole may be the only thing that allows for interstellar travel. Most of the people monitoring the transport machine were convinced that she never left Earth. The time dilation of the wormhole was so extreme that to everyone else, the trip was instantaneous. Yet to the traveler, 18 hours had gone by. Quite inconveniently, the recording devices she brought with her only recorded static. Because it seemed to the scientists that she had not gone anywhere, they doubted her story of having traveled through the wormhole. Later, two government officials confided that it was interesting that the static they had recorded lasted exactly 18 hours.

The movie ends with the words "For Carl" on the screen. How we love Carl Sagan!

New equations are needed

Perhaps the best argument for our not being the only civilization in the Universe lies simply in the numbers and statistical probabilities. If there are at least a hundred billion stars in the Milky Way galaxy, and at least a hundred billion galaxies in the Universe, is it not absurd to believe that we are the only civilization? Let's analyze this idea further.

The Drake Equation served well as a thought experiment for estimating the chances of life and civilization elsewhere in at least our galaxy. Today astrophysicists know a lot more than they did in the 1960s. What would an alternative, updated equation include? While recently there have been other proposed equations, I have laid out my own below that builds upon Drake's, and seeks to solve for how many intelligent civilizations have existed in the history of our galaxy.

My Suggested Alternate Equation with Estimates:

250 Billion: Number of stars in the Milky Way*

20%: Fraction of stars that live long enough to host life

40%: Fraction with relatively stable galactic orbits

70%: Fraction hosting planets of some kind

40%: Fraction with an Earth-sized planet in the habitable zone**

70%: Fraction of those planets that have land areas

40%: Fraction of planets with enough metals for technological needs

30%: Fraction of planets that can support life for at least a billion years

60%: Fraction that form complex life in a stable environment

30%: Fraction where life eventually becomes intelligent

60%: Fraction of intelligent creatures that physically can use tools

70%: Fraction of intelligent beings that build a civilization

80%: Fraction of civilizations that advance to radio communication technology or beyond

*One factor I did not include from the Drake Equation is the average number of new stars born in the Milky Way each year. I don't consider this as a very useful factor. A more pertinent factor than how

many newstars appear is the amount of stars that currently exist in the galaxy.

Life can last for billions of years, so the rate of new star production isgoing to be much less meaningful than the total number of stars thathave lived a good fraction of the age of the galaxy itself. Estimatesrange from 100-400 billion stars, so we will use 250 billion.

**Although most experts put this percentage at 20%, there is goodreasons to think it will be higher. With the latest telescopes and othernew technology, we are quickly coming to understand that Earth- andsuper-Earth-sized worlds are among the most common planets. We'refinding multi-planet systems all the time now, and many with planets inthe habitable zone.

So here's our equation in fraction form:

250 billion x .2 x .4 x .7 x .4 x .7 x .4 x .3 x .6 x .3 x .6 x .7 x .8 = 28,449,792

This equation suggests that at some point in the Milky Way's historyand near future, approximately 29 million civilizations with radiotechnology should be produced.

Now, if we multiply the amount of civilizations by the average time acivilization with radio communication survives, we will get the totalamount of years during which such civilizations will exist.

How long do radio-communicating civilizations last, beforedestroying themselves, setting themselves back in technology througherror (or on purpose) or by some natural event? If we base the averageon how long humanity has been using radio technology (about 100years) and how many times humanity has already risked catastrophicdisaster, I think 200 years is generous.

29,000,000 x 200 = 5.7 billion years of total existence time for all civilizations.

Let us now try to nail down when life could have first possibly arisenin the galaxy. Remember that the first generation of stars were devoidof heavy metals, and thus no orbiting planets would have been aroundthose stars. It would take a few generations of stars, and a few hundredmillion years, before planet formation could begin in earnest. Add tothat the time it takes for life to evolve from a single-celled organism to acivilization with radio technology, and we can estimate that such acivilization could not have existed before the Universe was about 5billion years old.

Deduct this first 5 billion years from the age of the Universe, 13.8billion years, and we know that our 29 million civilizations existed in thelast, roughly, 9 billion years.

If all civilizations lived during this period and at spread out times,with no two existing at the same time, then the maximum amount oftime over which they existed is 5.7 billion years.

We then take 9 billion years – 5.7 billion years = 3.3 billion years, minimum, when there were not technological civilizations.

How many such civilizations existed per year then?

5.7 billion / 9 billion = .6 civilizations with radio technology per year, on average, and at a maximum with that average.

There is more time than civilizations in existence, including if no two civilizations ever exist at the same time. This should exquisitely highlight the problem of two civilizations ever meeting each other.

These are just some of the conclusions we can come to, based onmy equation. You might want to create your own equation with updatedstatistics of the factors as they come about. Here are some otherfactors that would help refine our search:

- Fraction of stars with minimal flaring (referring to the star's stability)
- Fraction of habitable planets with a low orbital eccentricity (how circular the orbit is)
- Fraction of planets with a stable tilt (stabilized with a large moon)
- Fraction of planets with a sufficiently thin and oxygenated atmosphere

- Fraction of planets with the ability to recycle its atmosphere (through plate tectonics)

- How long it takes on average for at least single celled life to appear after a planet's formation (on Earth this was at least a billion years)

If 29 million intelligent civilizations spread out over the vastness of space and time doesn't sound like a lot, there is a silver lining for those who are keen to make contact with aliens (I am one of these people, as I am optimistic that they will not be hostile). The equation doesn't emphasize rarer civilizations with more advanced forms of communication that may last for thousands of years or longer, and spread throughout the galaxy like in Contact. There are those that should be able to expand beyond the constraints others find themselves in. If just a few can do this, then when we do detect a signal, it will likely be from one of these very rare, long-lived, and probably AI-based civilizations.

More good news for those who hope to make contact with aliens: the timeframe of when civilizations appear can be further narrowed to the last few billion years. There has been a bell curve in the rate of star formation, and we're past the peak of the curve. The curve peaked when the most sun-like stars existed in their mid to late stages of life. We know this because the rate of star formation in recent deep time is much lower than it used to be. If the rate were as high as it was billions of years ago, we would expect to see far younger stars compared to older stars, and we do not. It is likely that the rate of the evolution of civilizations with radio technology will mirror this bell curve, peaking a few billion years after the peak of star formation.

This civilization bell curve may play out like a thunderstorm. All of the elements for a heavy downpour begin to form, perhaps starting with a few sprinkles as the winds pick up speed. Even though the conditions are ripe for a heavy rain, it doesn't arrive for a while. The rain is light at first, but then it begins to come down in sheets. The downpour may last for just a few minutes, or perhaps for many hours, but eventually it tapers off back to a sprinkle of droplets, and then ceases altogether. This may be the historical picture of the rise and fall of civilizations in the galaxy. Humanity might be at the start of the tempest, and we are one of the first few drops before the downpour begins in earnest.

CHAPTER 7: ONE AMONG MANY

Possibly for the first time since the Universe began, matter and energy have come together in such a way as to be able to ask the ultimate questions about its existence: "What am I? How did I get here? Am I alone?"

Humanity lives in a unique moment in history. No civilization on Earth before us has experienced existence in quite the same way. The Mayans, Norte Chico, and Olmec never had our level of education, medicine, and security, not to mention the endless variety of entertainment options at the push of a button. If we could go back in time and experience what the lives of individuals in those early civilizations were like, we would probably be quite content to continue in the present with our air-conditioned homes and indoor plumbing. Each one of us is a unique thread of existence woven into a vast tapestry called civilization. This tapestry of humanity tells the story of an entire species' monumental effort to understand and explore its place in the cosmos. All that we have ever learned is contained on this single planet in computer archives, shelved in vast libraries, painted on ancient cave walls, and shared through stories passed down from one generation to the next. This knowledge is worth preserving for future generations of explorers and great thinkers.

We have a duty to all who came before us to act now to counter the threat of countless events that would guarantee our swift destruction, and erase all of our great history. To ensure that as many of those threats as possible are mitigated, we need to keep developing new technologies, secure the world's infrastructure, and educate the public in science. The dinosaurs didn't stand a chance against the asteroid that struck them. Humans also almost went extinct before – some say we got down to just 40 breeding pairs – after the supervolcano Toba erupted 72,000 years ago.

In the past, humanity didn't have the capabilities to prevent or dodge these calamities. Now we do. A diversified residence in the Universe is the ultimate solution to not only humanity's quest for survival, but also our ability to expand our experiences. Residing on multiple worlds would significantly reduce the risk of any single event wiping out everything we have created in one catastrophic blow. The greatest realization of the lofty goal of colonizing space is that it is entirely possible to make happen. We only need the will and focus to get it done.

Are there alien civilizations that have transcended the struggles that are a part of being an evolving species, and have successfully expanded beyond their home planet? We could learn from them, and perhaps they could learn a few things from us in return. The seeming emptiness of space would not be so empty if we knew of each other. If there are indeed other civilizations out there, it would be smart to present humanity in the best possible way. The scientists leading our quest into space are generally some of the brightest, most noble humans that can represent our ideals.

Although we may yearn for companions in the cosmos, we would be wise to not trust them too hastily. We could do without a stellar "frenemy" – or as they call it in French, faux ami – false friend. Wolves in sheep's clothing might catch us by surprise; we don't want to end up like the fools in The Twilight Zone's episode "To Serve Man." Feeling lonely might be an unfortunate consequence of being alone, but we might thank the heavens for the rather peaceful rapport with outer space which we have now. Mieux vaut être seul que mal accompagné, a French proverb, translates "It's better to be alone than in bad company."

If humanity one day ventures out to explore other star systems, only to discover worlds in ruin that once hosted thriving civilizations, then we should pay tribute to those civilizations and ensure they are remembered. Whatever evidence we can gather of their existence must be studied. And where an extinct civilization is found, a monument should be created there that preserves their identity and way of life, so that future generations can learn from their achievements, and their mistakes.

If dead worlds are indeed all that exists in our cosmic neighborhood, then perhaps we will have to adjust our hopes that a civilization could last for eons, and that we will ever be able to share our experiences with another intelligent species. We might have to accept that a more realistic goal for all civilizations is merely to live well and discover what they can, alone, in the time available to them.

Fate may yet deal us this most dire of cards as we attempt a journey to the stars. Someday humanity might fall back to a primitive society here on Earth, perhaps forever. If we at least did our best to establish a unified civilization that reached as far as it could into the depths of space, then maybe that is all that counts. Perhaps it will be some alien visitors eons from now that memorialize our once great civilization. They may honor the efforts humanity made to better itself and reach other sentient creatures that were indeed there, but just out of reach.

The ultimate quest then may not be to push forever forward one's own potential, but to learn about and remember the dignity of others. It is my hope then that we will be remembered well.

FURTHER READING

There are many topics in this book that draw upon knowledge fromcountless scientists and great thinkers. I encourage further reading intoall of them. Here are a few related books that will give you an evenbigger picture of our place in the cosmos:

Baggott, Jim. Origins. Oxford University Press, 2015.

Bartusiak, Marcia. The Day We Found the Universe. NY: Vintage, 2010.

Boyle, Godfrey. Renewable Energy. Oxford University Press, 2012.

Briggs, Roger P. Journey to Civilization. Collin Foundation Press, 2013.

Brown, Harris. The Challenge of Man's Future. Penguin Books, 1956.

Clark, Ronald. Einstein: The Life and Times. NY: Avon, 1984.

Davies, Paul. The Eerie Silence. UK: Penguin Books

Dawkins, Richard. The Greatest Show on Earth. Free Press, 2010.

Dawkins, Richard. The Selfish Gene. Oxford University Press, 2006.

Fichman, Frederick. The SETI Trilogy. Frederick Fichman, 2014.

Greene, Brian. Fabric of the Cosmos. First Vintage Books, 2005.

Greene, Brian. The Elegant Universe. NY: W.W. Norton & Company, 2003.

Greene, Brian. The Hidden Reality. NY: Random House, 2011.

Hawking, Stephen. A Briefer History of Time. NY: Bantam Dell, 2005.

Hawking, Stephen. The Grand Design. NY: Bantam Books, 2010.

Hawking, Stephen. The Universe in a Nutshell. A Bantam Book, 2001.

Krauss, Lawrence M. A Universe From Nothing. NY: Free Press, 2012.

Maloof, F. Michael. A Nation Forsaken. WND Books, 2013.

Morris, Simon Conway. The Runes of Evolution. Templeton Press, 2015.

Nolan, Christopher. The Science of Interstellar. W.W. Norton & Comp., 2014.

Nye, Bill. Unstoppable. NY: St. Martin's Press, 2015.

Sagan, Carl. Cosmos. Sagan Productions, Inc., 1980.

Sasselov, Dimitar. The Life of Super-Earths. Basic Books, 2012.

Savage, Marshall T. The Millennial Project. Empyrean Pub, 1993.

Schrodinger, Erwin. What Is Life? Cambridge, Eng.: Canto, 2000.

Stevenson, David S. Under a Crimson Sun. Springer, 2013.

Tyson, Neil deGrasse. Origins. NY: W.W. Norton & Company, 2014.

Ward, Peter and Brownlee, Donald E., Rare Earth: Why Complex Life Is Uncommon in the Universe, 2000.

Weir, Andy. The Martian. Random House, 2014.

Yavar Abbas. Earth: Making of a Planet. National Geographic Channel, 2011.

www.ingramcontent.com/pod-product-compliance
Lightning Source LLC
Chambersburg PA
CBHW040222220526
45473CB00001B/81

IGNITE INNOVATE LEAD

Unlock Your Creativity, Unleash The Entrepreneur Within, Cultivate A Growth Mindset and Transform Your Potential into Thriving Business

PRIYARATAN SHARMA

Copyright © 2024 Priyaratan Sharma

All Rights Reserved.

This book has been self-published with all reasonable efforts taken to make the material error-free by the author.

No part of this book shall be used or reproduced in any manner whatsoever without written permission from the author, except in the case of brief quotations embodied in critical articles and reviews.